Summary Contents

Chapter One: Understanding Emergencies and Disasters 1

Chapter Two: Natural Events 9

Chapter Three: Non-Natural Events 51

Chapter Four: The Basic Stages of Planning 87

Chapter Five: Creating the Emergency Action Plan 111

Chapter Six: Practice Makes Perfect: Training and Drills 137

Chapter Seven: Post-event Restoration 171

Chapter Eight: Guidelines for Emergency Mitigation 201

Appendix 211

Index 301

Contents

List of Figures xiii
Foreword xv

Chapter One: Understanding Emergencies and Disasters 1

Introduction: What Exactly Is an Emergency? 1
Natural Events 2
Non-Natural Events 2
Billion-Dollar U.S. Weather Disasters 3
The NOAA Weather Radio 4
Disaster Warning 4
FEMA's Impact 6
The Disaster Aid Process and Programs 6
Summary 7

Chapter Two: Natural Events 9

Blizzards and Avalanches 9
Facts and Figures 9
What to Do if Faced with a Blizzard or Avalanche 11
Understanding the Terminology 13
Winter Storm Watch 13
Winter Storm Warning 13
Blizzard Warning 13
Travelers' Advisory 13
Drought 13
Facts and Figures 13

What to Do if Faced with a Drought 14
Earthquakes 18
Facts and Figures 18
What to Do if Faced with an Earthquake 20
Fire 23
Facts and Figures 23
What to Do if Faced with a Fire 24
Floods 26
Facts and Figures 26
Understanding the Terminology 27
Flood Watch 27
Flood Warning 27
Flash Flood Watch 27
Flash Flood Warning 27
What to Do if Faced with a Flood 28
Hurricanes 29
Facts and Figures 29
Understanding the Terminology 32
Tropical Storm Watch 32
Tropical Storm Warning 32
Hurricane Watch 32
Hurricane Warning 32
What to Do if Faced with a Hurricane 33
Tornadoes 39
Facts and Figures 39
Understanding the Terminology 39
Tornado Watch 39
Tornado Warning 39
What to Do if Faced with a Tornado 41
Tsunamis 42
Facts and Figures 42
What to Do if faced with a Tsunami 43
Volcanoes 46
Facts and Figures 46
What to Do if Faced with a Volcano 47
Summary 49

Chapter Three: Non-Natural Events 51
Bomb Threat 51
Facts and Figures 51
What to Do if Faced with a Bomb Threat 53
The Bomb Threat Checklist 54
Whether to Evacuate 54
The Bomb Search 56
Chemical and Biological Concerns 61
Facts and Figures 61
What to Do if Faced with a Chemical or Biological Threat 62
Chemical Spills and Contamination 66
Facts and Figures 66
What to Do if Faced with a Chemical Spill 68
Civil Disturbance and Demonstrations 70
Facts and Figures 70
What to Do if Faced with a Riot or Demonstration 71
Loss of Utility 73
Facts and Figures 73
What to Do if Faced with Loss of Utility 74
Medical Emergencies 74
Facts and Figures 74
What to Do if Faced with a Medical Emergency 75
Nuclear Threat and Exposure 77
Facts and Figures 77
What to Do if Faced with Nuclear Exposure 78
Structural Collapse 80
Facts and Figures 80
What to Do if Faced with a Structural Collapse 81
Workplace Violence 82
Facts and Figures 82
What to Do if Faced with Workplace Violence 83
The Greatest Disaster That Never Happened 85
Summary 85

Chapter Four: The Basic Stages of Planning 87
Where Do I Start? 87

Breaking the Large Task into Bite-Sized Pieces 87
Determining Your Objectives and Scope 87
Identifying Your Resources 88
Understanding the Costs Involved 88
Presenting Your Case to Management 89
Justifying the Cost 89
Consequences Versus Rewards 89
The Mission Statement 90
Establishing Authority and Chain of Command 92
Understanding Your Business 92
Identifying Core Business 92
Cost of Business Interruption 93
Understanding Your Property 94
Leased Versus Owned 94
Geographic Location 95
Building Use 97
Understanding Your Risk 99
The Risk Assessment Matrix 99
Liability Versus Probability Versus Cost 100
Online Hazard Maps and Databases 101
Understanding Your Responsibilities 101
OSHA Regulations 101
Signage 102
Material Safety Data Sheets 103
Americans with Disabilities Act 105
Summary 109

Chapter Five: Creating the Emergency Action Plan ... 111

The Emergency Action Plan 111
What Does It Do? 111
What Does It Contain? 112
Who Gets a Copy? 113
The Emergency Management Team 114
Identifying Team Members and Responsibilities 114
Internal and External Support 115

Identifying Vendors and Contractors 116
Stocking Emergency Supplies 118
Blueprints and As-Builts 121
Critical Building Information 122
    Fire Extinguishers and Means of Egress 123
    Fire Suppression Systems 123
Pre-Planning and the Urban Search and Rescue Grid 125
Alarm Systems 126
    Understanding Your System 126
    Regular Testing 127
Hot Sites, Cold Sites, and Contingency Centers 128
The Command Center 129
Off-site Storage 130
    Data Security 130
Lack of Electricity Doesn't Mean "Go Home" 131
    Identifying Mission Critical Functions 131
    Uninterruptible Power Supply 132
    Emergency Generators 132
Summary 136

Chapter Six: Practice Makes Perfect: Training and Drills 137

So You Have a Plan. Now What? 137
Rolling Out the Emergency Action Plan 138
Training the Emergency Action Team 139
Training the Employees 140
Fostering Employee Support and Enthusiasm 142
Getting Everybody Out Safely 143
    Evacuation Routes and Drills 143
    Minimizing Interruptions and Down Time 146
    Alerting the Building Occupants 147
    Floor Captains and Brigades 150
    Shut Down Procedures 150
Assisting Those with Special Needs 152
    Use of Elevators for Exit 153
    Evacuation Assistance Devices 154

Areas of Refuge 155
The Staging Area 156
Accounting for Employees and Visitors 156
The Chain of Command 157
A Special Note for Those Who Refuse to Participate 159
Getting Back to Work 159
After the Drill: Evaluating Performance 161
Basic First Aid and Medical Care 162
The First Aid Kit 162
Burn Victims 164
Chemical Exposure 165
Broken Bones 166
Defibrillator 166
CPR 168
Summary 169

Chapter Seven: Post-Event Restoration 171
After the Event 171
Assessing the Damage 173
Insurance and Salvage Decisions 173
The Restoration Phase 175
Logistics 175
The Command Center 178
The Emergency Account Number 179
Catering 182
Providing Sleeping Accommodations 182
Environmental Issues 183
Important Asbestos Definitions 184
Asbestos Containing Material (ACM) 184
Presumed Asbestos Containing Material (PACM) 184
Surfacing Material 184
Thermal System Insulation (TSI) 184
Assisting Affected Employees 185
Don't Forget about Safety 186
Now Is Not the Time to Let Your Guard Down: Security 190
Monitoring Restoration Progress 191

Section	Page
Keeping the Lines of Communication Open	193
Communication with the Media	194
Communication with Employees	195
Communication with Families	196
Communication with Insurance Providers	196
Communication with Government Agencies	196
Communication with Customers	196
Sample Press Releases	197
Summary	198

Chapter Eight: Guidelines for Emergency Mitigation 201

Section	Page
Design Guidelines	201
Interior—Limited Access	201
Interior—Public Access	202
Point of Entry	202
Exterior—Building Perimeter	202
Exterior—Parking and Other Off-site Areas	202
Windows and Doors: The Weakest Link	204
Protective Bollards, Planters, and Green Space	205
Virtual Reality Software and Related Technology	206
Post-Failure Analysis	206
Building Codes	207
Building Performance Assessment Team	207
Summary	210

Appendix 211

Section	Page
Emergency Action Plan	213
Helpful Websites	290
Where Else Can I Find Assistance?	290

Index 301

Figures

Figure 1-1. Billion-Dollar U.S. Weather Disasters Since 1980 3
Figure 2-1. Heat Index 15
Figure 2-2. U.S. Drought Monitor 16
Figure 2-3. Seismic Hazard Map 20
Figure 2-4. Storm Activities List: Before the Storm 34
Figure 2-5. Storm Activities List: After the Storm 35
Figure 2-6. Contractor Availability Form 36
Figure 2-7. Fujita Tornado Categories 40
Figure 3-1. Bomb Threat Checklist for Phone 55
Figure 3-2. Search Procedures Step 1 58
Figure 3-3. Search Procedures Step 2 58
Figure 3-4. Search Procedures Step 3 59
Figure 3-5. Search Procedures Step 4 59
Figure 3-6. Search Procedures Step 5 60
Figure 4-1. Risk Anaylsis Matrix 101
Figure 4-2. Example of a Chemical Inventory List 105
Figure 5-1. Storm Preparation Inventory List 119
Figure 6-1. Location Key 151
Figure 7-1. Correspondence Log 180
Figure A-1. Contractor Availability Form 218
Figure A-2. Storm Preparations List 220
Figure A-3. Pre-Storm Activities List 252
Figure A-4. Post-Storm Activities List 253
Figure A-5. ATF Search and Sweep Procedures 273
Figure A-6. Risk Assessment Matrix 276
Figure A-7. Correspndence Log 277
Figure A-8. OSHA 300 Forms and Instructions 278

Foreword

Where were you on September 11, 2001?

Without a doubt, each of us will remember exactly where we were and what we were doing on that fateful morning. I was in New York City conducting a two-day building management seminar for American Management Association. On Monday, September 10, I opened the seminar by giving a brief preview of the topics we would discuss during the next two days. When I came to the subject of building security and disaster planning, I made my point by saying

> In this audience, you understand this threat better than most. In 1993, the World Trade Center just down the block was bombed by terrorists. Trust me when I tell you—it will happen again—and not only will the Trade Center be targeted, but a highly visible amusement park such as Disney, or the Super Bowl, or the upcoming Olympics in Salt Lake City will be targeted as well. Regarding terrorism, America has been on a short scholarship and that will change. When it does, the way we do business will change forever.

Sadly, the very next day these words became eerily prophetic. Of course, nobody could have predicted the depth of the destruction and sorrow we witnessed that day. Indeed we have been robbed of our innocence. Not only has our business changed but also our very lives and the values we hold dear have been irreversibly altered.

It is with this loss and the resulting concerns that I decided to author this book. Now more than ever, planning for disasters is paramount. We have been exposed, even forced to examine our own disaster planning and recovery plans, and typically, we don't like what we find. Surprisingly, only a few resources on this subject exist, even fewer with realistic and practical value.

Understand, however, that terrorist activities represent a small number of disaster events and are—thankfully—rare. More likely, you will be faced with storms, floods, hurricanes, tornadoes, earthquakes, losses of utility, and other such events. This book has been created to assist you and your organization in planning for and recovering from all events that fall into the category of emergency or disaster.

It is my sincere desire that this book provide you with the guidance you seek. Business continuity, productivity, and—most importantly—human lives are at stake. We have a responsibility to learn from past events, including those which occurred on September 11, 2001. In doing so, we can assure that the brave people who lost their lives did not die in vain.

About the Author

David A. Casavant, CFM, is the president of the Carlyle Consulting Group, Inc., a business consulting firm specializing in emergency planning and recovery for private and public industry. David is an active member of both the International Facility Managers Association (IFMA) and the Building Owners and Managers Association (BOMA). Mr. Casavant is an authorized OSHA trainer and a BOMI-approved instructor. To learn more about the Carlyle Consulting Group and the services it provides, visit the corporate website at www.carlyleconsultants.com.

Understanding Emergencies and Disasters

Introduction: What Exactly Is an Emergency?

Let's start with the basics. What exactly is an emergency? Is an emergency the same as a disaster? Good question. These two terms, although often used interchangeably, mean something quite different. An emergency can be defined as any event that negatively impacts an organization's production, safety, or personnel. That's a wide-open definition that encompasses many events.

The caution is to not view an emergency as a disaster. A disaster describes an emergency that has devolved into something with much greater impact. An emergency imposes a negative impact while a disaster's impact is both negative and severe. By concentrating on disasters, we overlook the more common and often more costly emergencies that affect us all. Disaster planning is fine and quite necessary; however, the purpose of this book is to identify any event that might cause disruption in your organization.

When we think of disasters, we tend to envision naturally occurring events such as hurricanes and tornadoes. We'd be remiss and sadly unprepared if we did not discuss all emergencies including non-natural emergencies.

Another point worth making is that the terms emergency and disaster are very elastic terms. What might be considered a disaster for one organization might be a mere inconvenience for another. For this reason, this book will usually use emergency to describe disruptive events. Whether the emergency is a disaster for your organization is dependant upon many unique variables.

Interestingly, all disasters start out as emergencies. It is not until the event has transpired that the emergency becomes viewed as a disaster. Think about large-scale emergencies such as Louisiana flooding in 1995. This single event brought up to 25 inches of rain to New Orleans in a five-day period and caused nearly five million dollars in damage. During the course of the event, it was considered a definite emergency; some people even called it a disaster. However, federal disaster assistance was not available until the president surveyed the damage and declared it a disaster. More than 90 percent of the floods that occur, although quite disruptive, are never declared disasters. Again, to focus on disasters would be a disservice.

The intent is not to get into a debate over words; that would be a frivolous use of time and energy. The fact remains: emergencies occur much more frequently than do disasters; therefore, an emergency is more likely to affect you and your business. While planning for emergencies, we are encouraged to ask the question "What is the worst that can happen?" Planning for a worst-case scenario is essential; however, we need to spend quality time performing a true risk analysis by asking "What events are most likely to happen?"

Typically, emergency events can be classified into two separate groups, natural events and non-natural events. When considering possible emergencies, remember to think about all events that might occur. As you read this book, you'll learn that many events we believe may have little chance of occurring might actually befall us.

With proper planning, you can significantly reduce the impact of disaster or emergency events on your business. In many instances, with proper planning, you can even prevent the emergency from occurring in the first place.

Natural Events

- Fire
- Drought
- Hurricane
- Tornado
- Tidal wave
- Blizzard
- Avalanche

Non-Natural Events

- Loss of utility
- Loss of communications
- Arson
- Water leak
- Chemical spill
- Act of war
- Riot or demonstration
- Bomb threat
- Medical emergency
- Workplace violence
- Nuclear accident

Billion-Dollar U.S. Weather Disasters

The United States has endured 49 weather-related disasters over the past 22 years in which the overall damage has exceeded $1 billion per event. The National Climatic Data Center (NCDC) is responsible for monitoring and assessing both national and international climate. Additionally, the NCDC identifies weather trends and provides summaries and comparisons with a historical perspective. A very interesting web page is the "Extreme Weather and Climatic Events" page (available at <www.ncdc.noaa.gov/ol/reports/billionz.html>) which identifies each of these 49 weather-related disasters and their costs (see Figure 1-1). The events include wildfires, floods, ice storms, nor'easters, heat, drought, and hurricanes. This sobering database is a great tool for opening the eyes of those who place low priority on emergency planning activities.

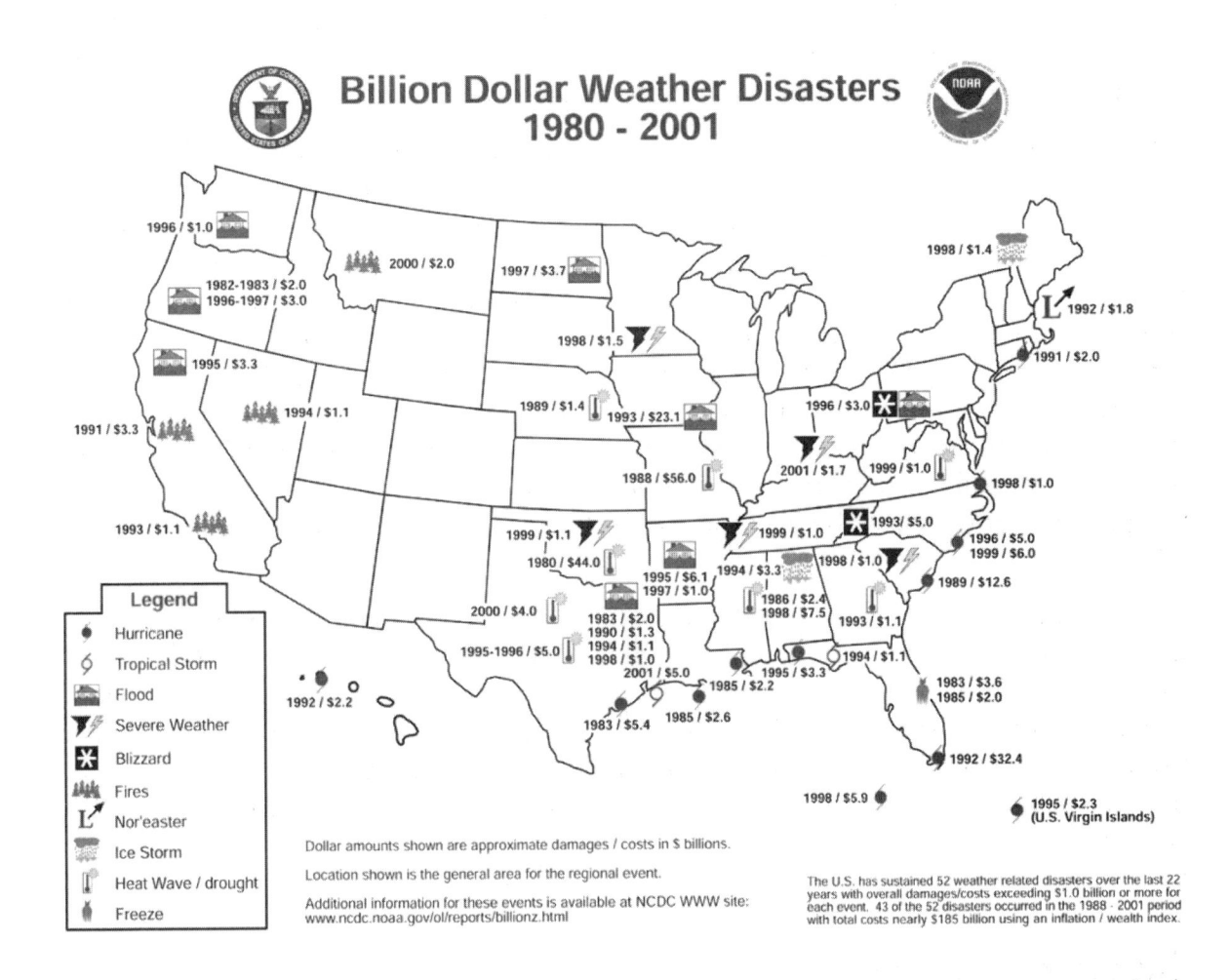

Figure 1-1. Billion-Dollar U.S. Weather Disasters Since 1980
Source: www.ncdc.noaa.gov/ol/reports/billionz.html

The NOAA Weather Radio

Your greatest source of information prior to or during any emergency event is the radio. During and after a major disaster, all radio stations drop their usual format and deliver helpful post-disaster information. It is critical to have a battery-powered radio to listen for storm warnings and advisories. Don't forget to supply extra batteries; you may be using the radio for quite some time after the storm has passed. Don't make the mistake of depending on electricity to power your radio; electricity may not be available.

The National Weather Service (NWS) broadcasts weather advisories and updates as received from the National Oceanic and Atmospheric Administration (NOAA). These broadcasts are available on NOAA weather radio, 24 hours a day. The advisories cover both natural events and non-natural events such as chemical spills and biological threats. NOAA weather radio has been created as an "all hazards" radio network, available to national, state, and local emergency management agencies. It is the most comprehensive emergency information source available to the general public.

Currently the NOAA radio network has more than 640 stations covering all 50 states, adjacent costal waters, the U.S. Virgin Islands, the U.S. Pacific territories, and Puerto Rico. The cost of the radio ranges from $25 to $200 or more, depending on the strength of the receiver and other features. Extra features include an alarm tone that allows you to keep the radio on but quiet unless a special signal is received by the national weather service. When this occurs, the radio will transmit the advisory. Another excellent feature, especially if you work in one area and live in another, is the SAME (Specific Area Message Encoding) feature. This option allows you to receive only those transmissions that relate to the areas important to you. Other features are available for those who are hearing or visually impaired. These options include attention-getting devices such as text printers, monitors, strobe lights, and even bed shakers. Purchase information for the NOAA weather radio can be found in the Appendix.

A limited number of National Weather Service broadcasts are available online in streaming audio. Access for these sites can be found through the National Weather Service web site at <www.nws.noaa.gov>. To listen to these web broadcasts, you'll need an audio player for your PC such as RealPlayer®. A free copy of RealPlayer can be downloaded from the National Weather Service website. If you do use the online version of the NWS broadcasts, remember that you'll be unable to receive them if you lose your Internet connection or electricity. These scenarios are very likely to happen, even if temporarily, during and after a severe weather emergency.

Disaster Warning

Effective emergency planning and accurate warning information coupled with speedy response have proven instrumental in saving lives and reducing losses. In particular, quick dissemination of warning information is essential to saving lives. For example, the use of Doppler radar has increased lead-time and accuracy of tornado warnings. Prior to 1990, the average lead-time for a tornado was two or three minutes and accuracy was low. The average lead-time for the years 1993–1999 was about nine minutes with an accuracy of about 58 percent. The projected lead times for 2000–2005 is expected to be 14 minutes with nearly 73 percent accuracy. In fact, lead times of twenty and even thirty minutes are not uncommon in 2002.

Even when a quick and accurate warning is given, those at risk may not receive the message, and this reveals a central weakness in the current warning system. For example, on March 27, 1994, a tornado killed 20 worshippers at a service at the UMC Goshen Church in northern Alabama. The warning was issued 12 minutes prior to the tornado strike, but because the victims were in church, they did not receive the warning. Evidently, nobody else in the surrounding areas had time to warn or thought to warn the church members. Even if a NOAA weather radio were available, in this case it would have been rendered useless because this region is one of the few not covered by the NOAA weather radio.

In another case, 42 people were killed when a series of strong tornadoes swept through east-central Florida on February 22–23, 1998. The National Weather Service issued 14 tornado warnings and these warnings were widely broadcast via news media, but most people were asleep and did not hear the warnings. In this scenario, a tone-alert NOAA weather radio could have awoken these people and saved their lives.

Effective disaster warning depends upon both a public and private partnership. Most emergency warnings are delivered by government agencies, but the information is usually disseminated by public avenues such as radio and television. In 1963, the Emergency Broadcast System (EBS) was established to provide a vehicle for warning the public about national disasters. In 1994 the Emergency Alert System (EAS) replaced the Emergency Broadcast System. The EAS is the national warning system designed to allow the president of the United States to address the nation in a reliable manner during and after a major national emergency.

At a national level, the EAS can only be activated by the president of the United States or his constitutional successor. After the president has used the system, it may be used by other federal agencies, such as the Federal Emergency Management Agency (FEMA). At a local level, the EAS can also be used by state and local agencies to release emergency notifications. In fact, the EAS is activated more than 100 times per month at state and local levels. These messages may be originated by the National Weather Service, police, governors, or emergency managers. The EAS will interrupt normal television programming and provide an appropriate verbal warning or "crawl" located at the bottom of the screen.

FEMA's Impact

A good record of major emergencies that have affected the country is the Federal Emergency Management Agency records. For the 1990-99 period, FEMA spent more than $25.4 billion for declared disasters and emergencies. This is in comparison to $3.9 billion (current dollars) in disaster aid for the 1980-89 period. This increase in spending can be attributed to a number of factors including public awareness of FEMA, increased population in trouble areas, and increased funding by the federal government.

During the 1990s, hurricanes and typhoons were the most costly weather-related events, for which FEMA has currently obligated more than $7.78 billion. Included in those costs were $2.5 billion for Hurricane Georges in 1998 and $1.8 billion for Hurricane Andrew in 1992. Flooding resulting from severe storms and other causes was the most frequently declared disaster. The 1994 Northridge earthquake in southern California stands out as FEMA's most dramatic and costly single disaster recovery effort, requiring nearly $7 billion in FEMA funding.

The Disaster Aid Process and Programs

First response to an emergency is the responsibility of the local and state government emergency services and various volunteer agencies. If the emergency is of large scale, the affected state's governor may request federal resources, which can be mobilized through the Federal Emergency Management Agency (FEMA). FEMA may provide search and rescue assistance, electrical power, food, water, shelter, and other basic human needs.

Usually, if the emergency has a short-term affect, FEMA assistance is not required; however the long-term recovery phase of a disaster can place severe financial strain on a local or state government. Even a large city with significant resources may be in need of federal assistance if the damage is severe. Of course, the infusion of federal funds requires the affected state to commit considerable funds as well. If the president determines the disaster to be of greater scope than the state can handle, a presidentially declared disaster is acknowledged and funding comes from the President's Disaster Relief Fund, which is managed by FEMA.

There exist two major categories of disaster aid:

- *Individual Assistance*—For damage to residences and businesses or personal property losses
- *Public Assistance*—For repair of infrastructure, public facilities, and debris removal.

In the event of an emergency event that the president officially declares a disaster, FEMA disaster workers will arrive and setup a central field office to coordinate the recovery effort. Shortly thereafter a toll-free telephone number will be established for use by affected residents and business owners. Additionally, disaster recovery centers will be opened where disaster victims can meet FEMA representatives to obtain information and apply for federal assistance.

Disaster aid to individuals generally falls into the following categories:

- *Housing* may be available for up to 18 months for displaced persons whose residences were heavily damaged or destroyed. Funding also can be provided for housing repairs and replacement of damaged items to make homes habitable.
- *Grants* are available to help meet other serious disaster-related expenses not covered by insurance. These expenses may include replacement of personal property; transportation; and medical, dental, and funeral expenses.
- *Low-interest loans* are available after a disaster for homeowners and renters from the U.S. Small Business Administration (SBA) to cover uninsured property losses. Loans may be for repair or replacement of homes, automobiles, clothing, or other damaged personal property. Loans are also available to businesses for property loss and economic injury.
- *Other aid programs* include crisis counseling; disaster-related unemployment assistance; legal aid; and assistance with income tax, Social Security benefits, and Veterans' benefits.

In addition to individual assistance, FEMA offers public assistance to assist state or local governments in rebuilding a community's damaged infrastructure. Typically, FEMA's public assistance programs pay for 75 percent of the approved project costs. Projects may include debris removal and repair of damaged public property, in addition to other public concerns.

Summary

Whether speaking of disasters or emergencies, businesses must be prepared.

Those businesses unprepared for emergency events will experience costly business interruption. More importantly, lives are at stake. History has shown that billions of dollars are lost each year due to emergency and disaster events and many organizations are ill prepared to survive an extended period of downtime.

Organizations exist that can offer assistance during and after an emergency event. In the event of a costly or long-term emergency, the governor of a state may petition the federal government for assistance. When this occurs, the Federal Emergency Management Agency quickly moves to provide assistance to individuals, businesses, and public entities.

2

Natural Events

As discussed previously, when we think of emergencies and disasters, we tend to envision naturally occurring events such as hurricanes and tornadoes. These natural events do tend to occur more frequently than non-natural events such as nuclear exposure and loss of utility. We'd be remiss and sadly unprepared if we did not discuss all emergencies including non-natural emergencies, but in this chapter we will focus on the natural events.

Blizzards and Avalanches

Facts and Figures

Each year winter storms disrupt our daily activities. At their worst, severe winter storms can result in blizzards and avalanches, which can cause death and property damage. In storms of lesser severity, transportation routes may be closed, communication may be cut off, and water pipes may freeze or even burst. The accumulation of snow and ice creates dangers for pedestrians and commuters alike. The cost of snow removal and the loss the business may have great negative economic impact on the area as well. Many areas in the United States are susceptible to severe winter weather. For instance, from the Middle Atlantic coast up to the New England coast, you might experience storms that originate off of the Carolina coast and move northward. These storms are commonly known as Nor'easters and can be very intense and include ice storms, strong winds, and flooding. The northeastern United States is also susceptible to severe winds originating from Canada. The Rocky Mountains tend to create severe weather as well, with cool air flowing eastward from the mountains meeting cool air from Canada and the Great Lakes. This places the Midwest at risk for winter storm events.

Even those areas such as southern California or southern Florida, which do not typically experience severe winter weather conditions, are at risk for winter dangers. Because of the rarity of severe weather in these more temperate areas, many people are not properly prepared. Weather dipping into the twenties or thirties may present a mere inconvenience in South Dakota, while in Florida it creates an emergency condition. The important point to understand is severe winter weather can occur in just about any region in the country; therefore, preparations should be made prior to the event.

Case Study 1: Great Appalachian Blizzard, 1996

The Great Appalachian Blizzard of 1996 began on Sunday, January 7, and didn't stop until Friday, January 12. What made the impact of this storm so great is the effect it had on the nation's capital. Just before the blizzard struck, the Republican Congress and the Democratic President squabbled over the federal budget for an extended period of time. Due to this political impasse, government offices were shut down for nearly a month into early January. Finally, both sides found a palatable medium, and thousands of federal workers were ready to come back to work on Monday, January 8. However, the extreme weather did not cooperate.

On Monday morning, the citizens of the District of Columbia awoke to a city buried beneath 17 to 21 inches of snow. The nation's capital was paralyzed. Other locations were hit even harder. Baltimore was sacked by two feet of snowfall, and nearly three feet of snow fell over Maryland's Frederick and western Loudoun Counties. Road crews worked hard to clear Monday's snow. As these road crews cleared the major arteries and began to tackle the secondary roads, a second storm blew through the District, dumping an additional three to five inches of snow. The snowplows assigned to clear the secondary and residential roads now had to go back and hit the main roads again.

Just as the population was beginning to sense relief and completion of this blizzard, a third storm struck on Friday, January 12, dumping another four to six inches over the metro area. By late Friday afternoon, the entire area was buried under two to three feet of snow. Because of this incredible week of severe weather, the government did not open its doors. It remained closed for most of the week while many schools and businesses gave up and closed their doors for the entire week.

Case Study 2: Northeast Ice Storm, 1998

The winter of 1998 began uneventfully enough. The northeastern United States was cold, but that was not unusual. However, during the week of January 5–9, the eastern United States and Canada was hit with a severe four-pronged weather event. A large storm system originating in Canada and traveling deep into the southern United States brought freezing rain, ice storms, flooding, and even tornadoes.

During a two-day period nearly 16 inches of rain fell in Tennessee and the Carolinas. The heavy rain destroyed over 700 homes and caused $35 million in damages in these two states alone. Tornadoes were spawned across the east and flooding was prevalent in many locations as well. Ice storms created another problem. In Canada, nearly three million customers were without electricity while over a half million U.S. customers were in the dark. The economic impact of this week of severe weather approached $3 billion for Canada and $1.4 billion for the United States. Flood- and storm-related deaths in the United States and Canada totaled 56.

What to Do if Faced with a Blizzard or Avalanche

Severe weather can be very dangerous and may impede your company's production. Because of this, you should have procedures in place to reduce or eliminate the negative effects of severe winter storms. For instance, a de-icing schedule and procedures should be written for your parking areas, loading dock, employee access areas, and other important locations. Create a process for removing snow from walkways. Timing is important; don't clear the snow too early or the pathways will be covered with snow before employees and visitors visit your buildings. In addition to de-icing, here are a few other proactive measures for effectively dealing with severe winter weather:

- Keep snow removal tools (shovels, salt, etc.) on hand.
- Create a rotating schedule for maintenance employees to be on call after hours in the event of an emergency.
- Consider using outside vendors for snow removal.
- Have a NOAA radio on site to monitor conditions.
- If you have a generator on site, remember to perform preventative maintenance on it, including startup and run.
- Beware of snow accumulation on roofs, which may become heavy and cause the roof to collapse.
- Check roof drains, gutters, catch basins, and other areas that can freeze or get clogged and impede the flow of melted snow.
- Winterize HVAC, fire sprinkler, and irrigation systems during the fall.
  - Lawn irrigation
    - Turn off the water to the irrigation system at main valve.
    - Turn the automatic irrigation controller to the "off" setting.
    - Turn on each of the valves to release pressure in the pipes.
    - Drain all of the water out of any irrigation components that might freeze.
    - Drain water out of the components and irrigation lines using compressed air.
  - Boilers
    - Drain idle equipment.
    - Check all service lines for freezing.
    - Elevate dead ends and low points.
    - Install heat tracing around control-line transmitter boxes and piping that carries water glass.

⚐ HVAC Cooling Coils
- Valve off coil and drain to level recommended by manufacturer.
- If an automatic freeze protection system is in use, test it for its proper use.
- Add a mixture of water and glycol, as recommended by manufacturer to coils where water cannot be drained.

Fire sprinkler systems freeze faster and therefore are at greater risk than standard plumbing pipe because the water doesn't flow unless there's a fire. In a dry pipe system, low points are susceptible to freezing due to the collection of condensation. Low points in a dry pipe sprinkler system where condensation can collect are susceptible to freezing. In wet pipe systems, the use of antifreeze can be a deterrent to freezing. Also, the size of the pipe plays a role in its likelihood to freeze. A smaller diameter pipe will freeze much more quickly than a larger diameter pipe. Remember, fire sprinkler pipes cannot be insulated or heated. If you utilize either of these methods you will defeat the purpose of the system or accidentally trigger the release of water.

Antifreeze loops or dry pipe applications are a simple solution for small unheated areas. It is recommended that you consult with your fire safety engineer or contractor.

> The best advice for keeping fire sprinkler systems operational during the winter months is to keep the heat on. It is recommend that all sprinklered rooms be maintained at 40 degrees Fahrenheit or warmer to avoid the possibility of pipes freezing. If you have concerns about the possibility of your sprinkler system failing as a result of cold weather, call your contractor and get professional advice.

Sometimes it will be advisable to let your building occupants leave work early so they can get home safely. Create a process for making this happen. How will you communicate this? Who will be required to remain? Who will shut down the building? Will anyone stay on site and monitor the site during the storm? If yes, will they be able to keep warm and fed? The time to create this process is before the severe weather sets in, not during.

Understanding the Terminology

Winter Storm Watch

Severe winter weather is possible within the next 36 hours.

Winter Storm Warning

Severe winter weather is expected within the next 24 hours.

Blizzard Warning

Severe winter weather with sustained winds of at least 35mph is expected. Expect near zero visibility.

Travelers' Advisory

Severe winter conditions may make driving difficult or dangerous.

Drought

Facts and Figures

Does a lack of rainfall actually constitute an emergency? Perhaps surprisingly, the answer can be yes. Between 1980 and 1999, the United States has endured 46 extreme weather events which each resulted in damage of $1 billion or greater. The cost of the 46 events exceeded $275 billion. Drought is responsible for nearly half of those losses, $120 billion. In addition to the economic losses, the toll on human life is large as well, exceeding ten thousand deaths as a result of drought and heat-related stress deaths. Failure to account for the possibility of drought would be a serious oversight!

Drought is one of those conditions that can affect just about any region of the country, even land that is surrounded by water. Drought is often accompanied by severe heat, which can complicate recovery as well. If the drought lasts a long time, vegetation will dry up and fires can easily be started. When drought conditions exist, wildfires can become difficult to control and extinguish. Drought can have a dramatic effect on the environment as well. Without a doubt, drought can cause serious interruption in both business and personal life.

Case Study 1: California's Seven-Year Drought, 1987–1994

The worst drought in 50 years affected not just California but 34 other states as well. Some areas had not seen rain for four years. Nearly half of Yellowstone National Park—

over two million acres—had burned as a result of the drought. Many homes were destroyed by the wildfires and billions of dollars in damage were assessed due to direct effects of the drought. In many areas mandatory water restrictions were enforced. Some communities even began to reward homeowners who removed the grass from their yards and planted vegetation that required little irrigation. This form of landscaping became known as "zeriscape" and produced a new industry. The drought even had an impact on the cost of electricity. Because of the reduction in hydroelectricity, a significant increase in the cost of electricity was absorbed by Californians.

Case Study 2: Southern Heat Wave, 1998

During the twelve-month period starting in September 1999 and ending in August 2000, the driest period was experienced in the southern United States since records began being kept in 1895. Only five months out of 28 had a positive "Z" index, a reading on the Palmer Z index which measures the moisture deviation from normal. In Florida, forestry officials estimate four million trees were lost in 2000 alone. Mandatory water restrictions were enacted in many Florida counties based on addresses. If your address, business or residential, was an even number, you were only allowed to irrigate your lawns on selected days for a limited period of time. Odd numbered addresses were permitted to irrigate their lawns on other specific days. Washing automobiles was strictly prohibited. Those breaking the drought law were subject to fines.

What to Do if Faced with a Drought

In a typical year, about 175 people die due to extreme heat. In an effort to save lives, the National Weather Service has developed the "Heat Index" (HI), an accurate measure of how hot it feels based on actual temperature plus the relative humidity. Think of it as wind chill factor in reverse. For example, in a very humid area such as Florida, a temperature of 94 degrees with 75 percent humidity actually feels like 124 degrees. Figure 2-1 shows the heat index.

> During the great drought of 1988, conditions were so bad in North Carolina that Governor Guy Hunt led a statewide prayer for rain, and—incredibly—it rained the very next day and continued for the next two weeks! This might not be the "official" solution to the drought problem, but in this case, it was effective nevertheless.

If, based on the heat index, the National Weather Service deems the heat excessive, they will issue a warning to the public. A common benchmark for excessive heat is a sustained daytime heat index equal to or greater than 105 degrees Fahrenheit, and a nighttime sustained temperature of 80 degrees F for two or more consecutive days. When this occurs, the National Weather Service broadcasts warnings via the NOAA weather radio.

Obviously, if the National Weather Service deems it necessary to inform the public of excessive heat, we as employers should do the same, especially when we have employees working outdoors.

Temperature	Relative Humidity (%)								
(°F)	90.0	80.0	70.0	60.0	50.0	40.0	30.0	20.0	10.0
65	65.6	64.7	63.8	62.8	61.9	60.9	60.0	59.1	58.1
70	71.6	70.7	69.8	68.8	67.9	66.9	66.0	65.1	64.1
75	79.7	76.7	75.8	74.8	73.9	72.9	72.0	71.1	70.1
80	88.2	85.9	84.2	82.8	81.6	80.4	79.0	77.4	76.1
85	101.4	97.0	93.3	90.3	87.7	85.5	83.5	81.6	79.6
90	119.3	112.0	105.8	100.5	96.1	92.3	89.2	86.5	84.2
95	141.8	131.1	121.7	113.6	106.7	100.9	96.1	92.2	89.2
100	168.7	154.0	140.9	129.5	119.6	111.2	104.2	98.7	94.4
105	200.0	180.7	163.4	148.1	134.7	123.2	113.6	105.8	100.0
110	235.6	211.2	189.1	169.4	151.9	136.8	124.1	113.7	105.8
115	275.3	245.4	218.0	193.3	171.3	152.1	135.8	122.3	111.9
120	319.1	283.1	250.0	219.9	192.9	169.1	148.7	131.6	118.2

Less than 80 is considered comfortable.
Greater than 90 is considered extreme.
Greater than 100 is considered hazardous.
Greater than 110 is considered dangerous.

Heat Index	Possible Heat Disorder
80°F–90°F	Fatigue possible with prolonged exposure and physical activity.
90°F–105°F	Sunstroke, heat cramps and heat exhaustion possible.
105°F–130°F	Sunstroke, heat cramps, and heat exhaustion likely, and heat stroke possible.
130°F or greater	Heat stroke highly likely with continued exposure.

Figure 2-1. Heat Index
Source: www.crh.noaa.gov/pub/heat.htm

The Drought Monitor website (Available at <http://www.drought.unl.edu/dm/monitor.html>) is the result of a partnership between federal and academic scientists. The two main federal partners are the U.S. Department of Agriculture (USDA) and the National Oceanic and Atmospheric Administration (NOAA). These two agencies have joined forces with the National Drought Mitigation Center of the University of Nebraska–Lincoln. Current drought information is gathered and released each Thursday at 8:30 a.m. (EST). A map (Figure 2-2) is posted and drought conditions are graphically represented using the following indicators:

D0 – Abnormally dry
D1 – Drought Moderate
D2 – Drought Severe
D3 – Drought Extreme
D4 – Drought Exceptional

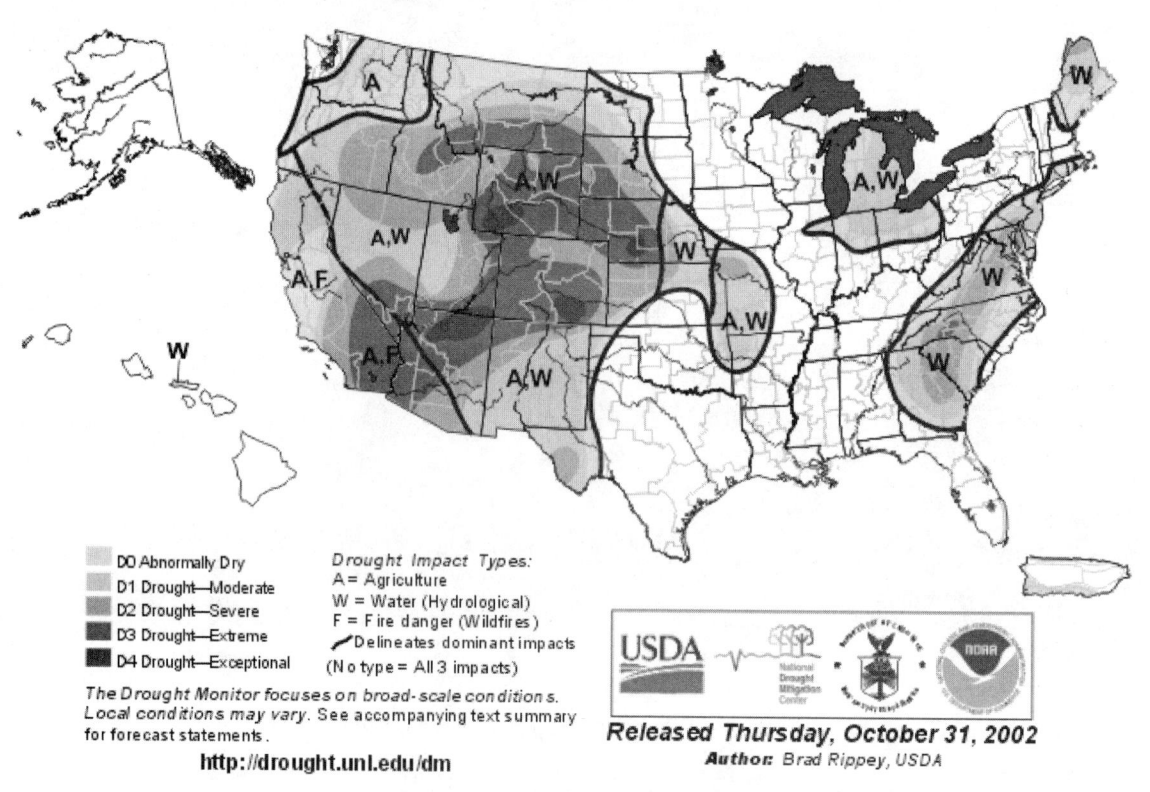

Figure 2-2. U.S. Drought Monitor

Also, regions that are susceptible to wildfire as a result of drought are identified. From this website you can view drought forecasts, historical data, and a written synopsis of the current conditions.

The United States does not have a cohesive national water policy. Instead, each state or local municipality has taken the lead in developing proactive mitigation and conservation policies. As recently as the U.S. drought of 1977, no state had a formal drought plan. In 2002, only ten states did *not* have a formal drought plan. Based on this fact, the first source of drought-related planning for businesses should be your state or local municipality. Here is a step-by-step method for developing an effective drought plan:

Step 1

Consult your local municipality or state for an established plan. If they do have a formal plan, use it to develop your plan in parallel.

Step 2

Conduct a historical drought analysis for your location. Do this by visiting the Drought Monitor website at <http://www.drought.unl.edu/dm/monitor.html>.

Step 3

Categorize the risk. If your historical analysis identifies a high probability of drought, proceed with the drought plan.

Step 4

Identify specific areas that are drought prone. This might include large areas of vegetation or property that borders on wooded areas. Non-irrigated properties might pose a cosmetic problem as well.

Step 5

Establish terms that identify the severity of the drought such as:
- Advisory
- Alert
- Emergency
- Ration

> The National Drought Mitigation Center (NDMC) was established in 1995 and is based in the School of Natural Resource Sciences at the University of Nebraska–Lincoln. The NDMC provides drought monitoring, planning, and mitigation. They also maintain a database of historical drought occurrences. They work with various governmental agencies to create drought policies and answer frequently asked questions. The NDMC conducts workshops and conferences that address the problems of drought. They are an excellent source of information for drought planning and recovery. Their website is <www.drought.unl.edu/>

These terms are more effective than numbers to identify the severity.

Step 6

Based on the severity and the term assigned, develop steps for each. For instance, when an "advisory" is announced, steps might include reduction of water use and a visual inspection of the premises to identify potential fire hazards.

Step 7

During and after the drought, be certain to evaluate what worked and what didn't work. Revise your drought plan accordingly.

Here are a few other ideas that may assist in your drought mitigation efforts:
- Always be aware of and obey the water rationing laws.
- Train employees in water conservation activities and set expectations for compliance.
- Trim back vegetation from the building and premises that could dry out and catch fire.

- Be aware of work that could cause fires such as welding, roofing, and grinding. Take precautions to protect against these potential fire triggers.
- Be aware of chemical storage and spills that could cause fires.
- Ensure that fire systems and fire extinguishers are working properly.
- Provide employees with training for proper fire control and building evacuation.

Earthquakes

Facts and Figures

When you think of earthquakes, chances are you think of California. True, the Pacific Coast seems to be particularly susceptible to earthquakes. Earthquakes are most likely to occur west of the Rocky Mountains, although forty states have been identified as having at least a moderate earthquake risk. Even areas not typically thought of as susceptible to earthquakes such as Boston, Massachusetts, and Charleston, South Carolina, have experienced seismic activity.

Earthquakes occur without warning as the earth's tectonic plates stress and shift. Although earthquakes typically last only seconds, they can inflict tremendous damage. At their worst and most violent, earthquakes will cause buildings and highways to collapse. Even if a total collapse does not occur, people can be injured by falling objects and flying glass. Additionally, downed power lines and ruptured gas lines add to the casualties. Fire sprinklers are often triggered by earthquakes, which results in even more property damage.

After the earthquake has occurred, it is not uncommon to experience multiple aftershocks. These aftershocks are particularly dangerous because they may occur days later when unsuspecting people are engaged in emergency recovery efforts.

One byproduct of an earthquake, especially in areas with hilly terrain, is landslides. As soil becomes unstable, buildings collapse and take other structures with them. Buildings located on the sides of slopes or near the foothills need to be especially aware of the potential danger.

Case Study 1: Loma Prieta, 1989

At 5:04 P.M. on October 17, 1989, an earthquake measuring 7.1 on the Richter scale rocked the California Coast. The epicenter of this powerful quake was approximately 60 miles south of San Francisco, but the greatest impact occurred in the heavily populated San Francisco/Oakland Bay areas. The shockwaves were felt as far south as San Diego and west to Nevada.

Many people will remember the horror of seeing an elevated section of Interstate 880 in Oakland that had collapsed onto the lower section below. The majority of the 62 fatalities caused by the Loma Prieta earthquake occurred in this location. Also notable was the postponement of the third game of the World Series that was to be held that evening at Candlestick Park in the Bay area.

Damage and economic impact was estimated at about $10 billion. Most of the damaged and destroyed structures were residential homes, but over 2,500 commercial buildings were damaged and 147 were totally destroyed. Sixty-two people died and over 3,700 were injured. Thankfully, these numbers were lower than they might have been considering the timing of the earthquake. Many people were just leaving work. If Candlestick Park, with its arriving fans, had been hit more intensely, one fears to think about the loss of life that may have occurred.

> A potential byproduct of an underwater earthquake is the tsunami. Tsunamis occur when large volumes of water are displaced due to a shifting of the ocean floor. The waves of a tsunami can travel at speeds of up to 600 miles per hour and reach heights of 100 feet!
>
> Two agencies are able to provide specific information about your risk:
>
> West Coast Tsunami Warning Center of NOAA
> (includes Alaska)
> 910 S. Felton Street
> Palmer AK 99645
> (907) 745-4212
>
> Pacific Tsunami Warning Center (Hawaii)
> 270 Fort Weaver Road
> Ewa Beach, HI 96706
> (808) 689-8207

Case Study 2: Northridge, California, 1994

The Loma Prieta earthquake of 1989 was fresh in the minds of many Californians when their worst fears were realized once again. At 4:31 a.m. on January 17, 1994, a quake measuring 6.8 on the Richter scale struck southern California. This earthquake became known as the Northridge earthquake. Although not as intense as the earlier Loma Prieta quake, the Northridge earthquake caused a greater economic impact. Thousands of aftershocks, many measuring 5.0 on the Richter scale, continued to rock the area, further destroying properties and making the recovery effort even more difficult.

The cost was high both in lives and property damage. Fifty-seven people lost their lives and over 8,700 were injured, 1,500 of those seriously. Well over 100,000 structures were damaged while nearly 4,000 of these structures were severely damaged and tagged for condemnation. Many properties were without electricity, water, or gas for an extended period of time. Fires started as a result of the earthquake, which added to the damage. In retail businesses and warehouses, large inventories were lost due to earthquake damage, fire, or water from fire sprinkler systems. All told, the economic impact of this 20-second event and its aftershocks reached an estimated $20 billion.

Although the cost of the quake was large, California is no stranger to the devastating effects of earthquakes, and certainly past experiences helped to reduce the negative effects and recovery time necessary. Many of the properties destroyed were between 30 and 40 years old, built prior to stricter building codes. Response teams were able to quickly provide relief and ascertain the extent of damage. Additionally the early pre-dawn timing of this quake no doubt preserved the lives of people who might have otherwise been traveling on roads and bridges that collapsed.

What to Do if Faced with an Earthquake

The first step is to determine the probability of an earthquake affecting your location. This can be accomplished by visiting the National Seismic Hazard Mapping Project at <http://geohazards.cr.usgs.gov/eq/pubmaps/US.pga.050.map.gif.> Many maps detailing earthquake probability similar to the one shown in Figure 2-3 are available at this site.

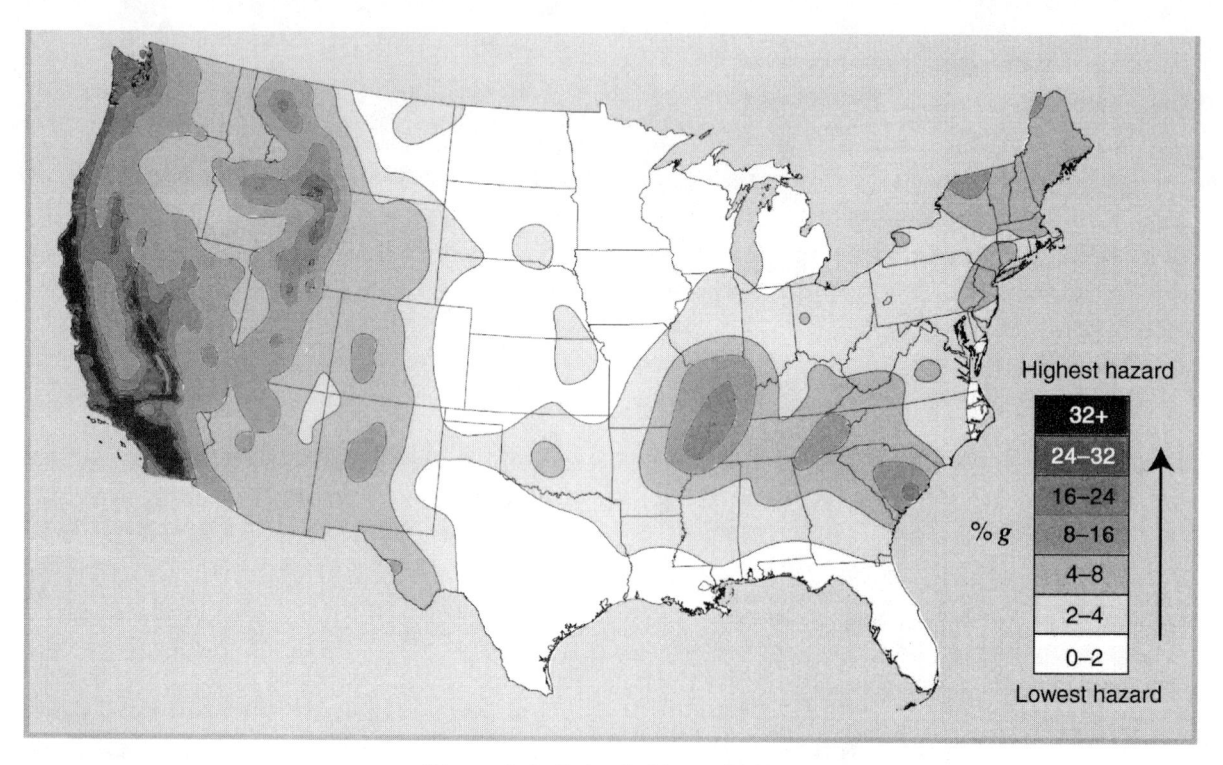

Figure 2-3. Seismic Hazard Map

If you are located in an area identified as 2 % g (peak acceleration due to gravity) or less, you have a relatively low probability of an earthquake occurrence and may elect not to spend time on this hazard. Conversely, if you are located in an area of 3 % g or more, you have a greater probability of an earthquake occurrence and should plan for earthquake emergencies.

Prior to an earthquake you should have a detailed plan for handling this emergency. Elements of this plan should include

- Safe locations in the building
- Danger locations in the building
- Practice drills
- Utility shut off locations
- Evacuation staging areas
- Communication procedures with employees and families

An emergency survival kit would include

- Non-perishable food
- Water
- Portable battery-powered radio
- Extra batteries
- Basic tools (for utility shut-off)
- Personal medicine
- General first aid kit
- Clothing
- Money

It is important to understand that immediately after an earthquake, all emergency personnel will be reacting to critical crisis. For this reason you should be able to be self sufficient for at least 72 hours. Keep this in mind when preparing your plan and survival kit.

During an earthquake it is imperative that you remain calm. The majority of injuries during an earthquake occur when people rush to enter or exit a building. If you are indoors, take refuge beneath a desk or table. If that is not possible, move toward an inside wall away from items that could fall and injure you such as bookcases, vending machines, lighting, and other fixtures.

If you are outdoors, move away from buildings, radio towers, light poles, and overhead utility lines. In a high rise building, stay off of elevators. If you are on an elevator when an earthquake occurs, stay clam. The elevator may quit working. If power is lost, most elevators will automatically go to the lowest level. Don't be surprised if fire alarms or fire sprinkler systems go off.

Because of subterranean damage that occurs during an earthquake, electrical utilities, gas lines, and water lines are often ruptured and rendered useless. Although uncomfortable, a lack of utility is a survivable event. A lack of water, however, is more than just a nuisance. As we know, water is essential to life. This must be considered when planning for an earthquake. The average person consumes between two to two and one-half quarts of water per day. If limited water exists, do not ration the water to amounts less than one quart per person per day. Instead, provide each person with adequate water and immediately find more water! Here are a few of the "hidden" sources of water:

- Water in hot water tank (20 to 60 gallons)
- Water in toilet flush tanks (not the bowl!)
- Ice cubes
- Other liquids such as juice, soda, and fruits
- Water pipes

After the earthquake local authorities may advise you to turn off the main water valves that serve your buildings. If this occurs, follow the instructions. It will prevent water from evacuating your water pipes. To capture the water in your pipes, turn on the faucet at the highest point in your building. This lets air into the system. Next, draw water from the faucet located in the lowest point of your building.

When all clean sources of water are exhausted, you must now turn your efforts toward water that is considered unclean. This water must be purified prior to drinking. To properly purify the water, here are a few guidelines:

- Strain the water. Use cloth or paper towels if a filtration system is not available.
- Boil the water for three to five minutes.
- If boiling is not feasible, use purification tablets.

As in many emergencies, many injuries occur as a result of the after-effects, not directly as a result of the emergency itself. Many problems are a result of improper food storage, preparation, and handling.

- Don't forget normal hygiene practices. Wash hands when preparing food.
- Use clean utensils and cutting surfaces.
- Keep food in covered containers.
- If in doubt about spoiled food, avoid it.

The first and most important responsibility after an emergency event is to assist injured people. Do not attempt to move seriously injured people unless their location places them in additional danger. The following items should be addressed as well:

- Check for gas and water leaks. Turn off supply or report to utility company immediately.
- Check for downed power lines. Report to utility company.
- Examine the building for obvious structural damage. Warn others.
- Tune your radio for further information.
- Cooperate fully with public safety officials.
- If possible stay off the roads. Emergency personnel should be given first priority.

⋏ Contact family and employer.
⋏ Be prepared for aftershocks.

Fire

Facts and Figures

A fire is started every 18 seconds on average. Public fire departments attend to over 1.7 million fires in a typical year. In 2000, fires caused over 4,000 deaths and over 22,000 injuries. In a typical year property damage runs into the hundreds of millions of dollars.

When discussing the danger of fires, two distinct types of fire should be represented. The first category is fire that occurs within a building. Such fires usually start by a secondary source such as faulty electrical wiring, carelessness, or flammables. The second and perhaps most often overlooked fire is the forest fire. The forest fire, or wildfire, starts outside of a building structure and is fueled by vegetation. If the wildfire is not controlled, it can quickly consume buildings as well. People start over 80 percent of the wildfires that occur. Lightning strikes usually cause the remainder. Each of these types of fires present their own unique challenges and solutions. We will examine case studies and consider possible mitigation steps for both types.

Case Study 1: Beverly Hills Supper Club, 1977

On the evening of May 28, 1977, a festive crowd of nearly 2,800 gathered at the Beverly Hills Super Club in Kentucky to enjoy an evening of good food, friends, and entertainment. At the time, no one could have predicted that 165 of those party seekers would not come out of the club alive. In fact, most in attendance didn't realize the severity of the fire until hours later. Customers were seen exiting the burning building with drinks in hand, wondering why the nuisance.

The most often cited reasons for the fire and its devastation were faulty aluminum wire, inadequate exits, and lack of minimum firewall-rated materials used in construction. Many of those who died were not burned or even covered in soot. They appeared as if asleep, dead due to smoke inhalation. The fire burned for five long hours before it was extinguished. One hundred sixty-five people were dead, and all that remained of the Beverly Hills Supper Club was the front façade and a few partially standing walls.

Case Study 2: Western United States Wild Fires, 2000

During the spring and summer months of 2000, nearly seven million acres of land in the western United States was burned. The magnitude of these fires can be blamed on the unusually dry, hot summer and the presence of strong winds. These winds helped to fuel the fires and increased the speed at which they expanded. Lightning storms without the benefit of rainfall appeared to be the root cause of most of the wildfires. Because of the severe drought gripping the west, much of the vegetation had dried up and provided the fires with additional fuel.

By the time the raging fires were extinguished and the economic impact was calculated, an estimated $2 billion was spent. Fortunately, no lives were reported lost; however, hundreds of people were injured, including many of those assisting in efforts to contain the fires.

What to Do if Faced with a Fire

The first step is to determine the probability of a fire. Although not necessarily predictable, the probability of a fire can be reduced. If you are located in an area with a history of wildfires or if you are located near dense brush or forests, you should consider your risk to be escalated. Additionally, if you are currently experiencing extended drought, this condition increases the risk of wild fires.

The Occupational Safety and Health Act requires that employers protect employees from fire and other emergencies. Part 1910 of the OSH Act addresses general industry, which covers most operating facilities. Specifically 1910.38 (b) (1-5) requires the following components:

- A list of all the major workplace fire hazards and their proper handling and storage requirements
- Potential ignition sources (i.e., welding, smoking)
- Type of fire protection available to control each hazard
- Names of personnel responsible for maintenance of equipment installed to prevent or control fires
- Names of personnel responsible for controlling ignition sources
- Maintenance plan for fire equipment or systems (i.e., inspections and certifications)
- Housekeeping procedures to ensure that work area is kept free from accumulation of flammable and combustible materials

In addition to these distinct requirements, OSHA requires employers to provide adequate training to employees. The training must apprise employees of the fire hazards of the materials and processes to which they are exposed. A written plan must be available for all employees to review.

> If you desire to view the current threat of wildfires, visit the U.S. Fire Safety's Fire Danger maps at <www.fs.fed.us>. This page will help you to determine possible impending threats to your location from wildfires.

If you utilize a "fire brigade," employees who are assigned to assist others during a fire, their responsibilities must be set out in writing and included in the fire emergency plan, and those employees must be trained in their responsibilities. Their responsibilities might include

- Providing first aid to those injured (as well as moving them to safety)
- Controlling spectators and those not involved in the fighting of the fire
- Meeting the fire department and giving them information as requested
- Ensuring that the alarm is given throughout the building and that all people, including those with disabilities, are able to evacuate the building
- Providing communication between various staging areas and central command
- Using fire extinguishers as needed
- Communicating information about missing people to the fire department
- Turning off gas lines, machinery, blowers, and other equipment that may make the fire more dangerous

If you have difficulty determining what might constitute a fire hazard, meet with your local fire department and ask for its assistance. The firefighters will be happy to visit your facility and assist in identifying processes or materials that could cause a fire. Additionally, they will identify chemicals or other substances that will fuel a fire or contaminate the building during a fire.

A water sprinkler system is the most effective method of controlling fires automatically. These systems can be expensive to install and are usually required in new commercial construction, but the peace of mind they provide and the reduction in insurance premiums are well worth the cost. Three types of automatic sprinkler systems exist:

- Wet Pipe System—Water is held in pipes that are located in the building. When heat from a fire is present, the seals in the sprinkler heads are ruptured and the water is able to flow and douse the fire.
- Dry Pipe System—Instead of water in the pipes, air is present. When a sprinkler head is ruptured, the air rushes out and water flows through the pipes and to the fire. This application is useful when the possibility of frozen water in the pipes exists.
- Deluge System—This system is similar to a dry pipe system except that some of the sprinkler heads remain in the open position to facilitate water flow in a chosen direction. Because some of the valves do not have a heat seal and remain open, this system can be operated manually or automatically.

Floods

Facts and Figures

Floods are one of the most common of all the natural hazards. Flooding creates some degree of risk for just about every community. Obviously, residing next to a river or lake increases the likelihood of a flood, but don't be fooled into discounting a flood if a body of water is not present nearby. Snowmelt, broken water pipes, breached dams, tsunamis, hurricanes, excessive rain, or even lack of permeable soil (i.e., parking lots) can trigger a flood.

Floods kill an average of 150 people per year and cause billions of dollars in damage. Approximately nine million people currently reside in identified floodplains. Every state in the country has a potential for flooding.

In addition to the actual danger created by a flood, additional danger exists from water contamination, electrocution, structural instability, and drowning. Based on the likelihood of a flood affecting your property, it would be wise to include flood risks in your emergency planning efforts.

Case Study 1: The Great Chicago Flood, 1992

One of the more unusual floods in the United States occurred in Chicago in an event that became known as the Great Chicago Flood. This flood was not the result of storms or snowmelt. It was the result of millions of gallons of water from the Chicago River pouring into the basements of over 500 buildings in Chicago's downtown Loop. The river water was able to enter the basements because a contractor driving pilings around the Kinzie Street Bridge penetrated an abandoned tunnel system, which once provided downtown buildings with coal during the early part of the century. The tunnel system was largely forgotten and unused except by utility workers who used the tunnel system to run cables and conduits. When the lining of the tunnel system was penetrated, water filled the tunnels, which in turn filled the basements connected to the intricate tunnel system.

During the early stages of the flooding, water began to rise at nearly four feet per hour.

The economic loss, including repairs, cleanup, and lost business exceeded $2 billion. This flood proved costly because the basement areas of office buildings

> ### Danger: Flash Floods
>
> According to FEMA, nearly half of all flash flood fatalities are automobile related. When a vehicle stalls in floodwaters, the water's momentum is transferred to the car. For each foot that water rises, 500 pounds of lateral force is applied to the car; however, the biggest factor is buoyancy. For each foot the water rises up the side of the car, the car displaces 1,500 pounds of water. In effect, the car weighs 1,500 pounds less for each foot the water rises. A depth of two feet of water will carry away most automobiles.

traditionally house the mechanical heart of the buildings. Vital and expensive equipment such as electrical centers, generators, elevator rooms, chillers, and boilers are housed in these basements. Thanks to the flooded basements, much of this equipment was rendered useless. Although no deaths were caused by this emergency, some 250,000 people were impacted by the flood. Hundreds of buildings were shut down temporarily, some up three weeks.

What made this emergency so frustrating is the fact that the tunnels did not flood until six months after the pilings were driven. Engineers had noted some unusual activity in the tunnel and even provided an estimate of $10,000 to make the needed repairs! Obviously, the needed repairs didn't happen in time and the results were disastrous.

Case Study 2: New Orleans Flood, 1995

During the spring of 1995, heavy thunderstorms drenched Louisiana and the Mississippi Delta. During this 16-day spree, parts of Louisiana recorded 12 to 17 inches of rain. A period of normalized weather extended for the next three weeks and then the severe rain began to fall again. In many areas the rain fell quickly and resulted in flash flooding. During the next three days, up to 26 inches of rain fell on portions of southeastern Louisiana and other coastal areas extending all the way to the Florida Panhandle. In just eight hours, over 15 inches of rain fell at the National Weather Service's Slidell office.

Severe flooding was reported in Slidell, which borders Lake Pontchartrain, causing many homes to be evacuated. Flooding continued to New Orleans which is particularly susceptible to flooding because it is, on average, six feet below sea level and is bordered by Lake Pontchartrain and the Mississippi River.

Understanding the Terminology

Flood Watch

Floodwaters are possible in the given area of the watch and during the time frame identified in the flood watch. Flood watches are usually issued for flooding that is expected to occur at least six hours after heavy rains have ended.

Flood Warning

Flooding is actually occurring or is imminent in the warning area.

Flash Flood Watch

Flash flooding is possible in the given area. Flash flood watches are usually issued for flooding that is expected to occur within six hours after heavy rains have ended.

Flash Flood Warning

Flash flooding is actually occurring or is imminent in the warning area.

Texas, Oklahoma, and Mississippi were also hit hard by the continued rainfall. Some areas received 25 inches of rainfall in a five-day period. Over 32 deaths were recorded and damages of over $5 billion were assessed.

What to Do if Faced with a Flood

The first prudent step is to determine the risk of flood in your area. Fortunately, good information is available to determine your risk exposure. Start by examining your surroundings. Properties located in valleys or on land that borders water is most susceptible. If a water dam is within 50 miles of your property, consider your risk to be elevated. The Federal Emergency Management Agency (FEMA) has mapped most of the floodplains and determined the risk for those areas. FEMA's Flood Insurance Study (FIS) contains a Flood Insurance Rate Map (FIRM), which will be key in identifying your risk. The rate map identifies Special Flood Hazard Areas (SFHA's), designates flood zones, and establishes Base Flood Elevations (BFE). The Base Flood Elevation is the elevation of water that results from a "100-year Flood."

> The Federal Emergency Management Agency (FEMA) has launched Project Impact, an initiative to assist at-risk communities across the county by mitigating the impact of disasters. FEMA has joined forces with state and local government, the private sector, volunteer groups, and community organizations to create a safer environment and to more quickly respond post disaster.

To get a copy of a Flood Insurance Rate Map, contact FEMA Map Service at (800) 358-9616 or on the internet at <www.fema.gov/maps>. A secondary source of flood risk information can be found at <www.esri.com/hazards>.

The next step is to examine the FIRM to determine your exposure. If the exposure is great, a specific flood plan should be created for your property.

When evaluating your exposure to flood emergencies, the 100-year flood zone is the standard benchmark. If you fall within this zone, you should plan for the inevitable flood emergency. The 100-year flood is often misunderstood. In reality, the 100-year flood has a one percent chance of occurring in any given year. Such a flood may, however, occur two years in a row or multiple times during a 100-year time frame.

> ESRI was founded in 1969 as Environmental Systems Research Institute. Today they are leaders in providing geographic information system (GIS) mapping software and information. FEMA and ESRI have formed a partnership aimed at providing multi-hazard maps and information to the public via the Internet. The information provided at <www.esri.com/hazards/> is intended to assist in building disaster-resistant communities across the country by sharing geographic knowledge about local hazards.

Your property insurance provider will have access to FEMA's flood zone maps. When evaluating insurance, remember most policies do not include damage that occurs as a result of floods. A separate flood policy must be purchased.

When a flood threatens, your best source of information is the NOAA radio. If you do not have a NOAA radio, watch the television or dial into a local radio station and stay tuned for more information and further instructions. Remember to have a battery-powered radio and additional fresh batteries because electrical service may be interrupted during the emergency.

Here are a few of the guidelines when in a flood zone:

- Keep adequate flood insurance coverage.
- Become familiar with local emergency agencies and their plans.
- Keep a supply of sandbags available.
- Have visqueen and lumber available for repairs.
- Move valuables to higher ground (or higher floors).
- Shutoff utilities including gas and electric.
- Disconnect electrical appliances before water rises.
- Beware of using fire (candle or cooking) because gas may escape from ruptured lines.
- Always watch for downed electrical lines (report immediately).
- Use water portable pumps with discharge hoses to remove water from building.
- Remember to properly dehumidify the building after cleanup.
- Consider generator power for cleanup.
- Fill the bathtub with water in case of service disruption.

> If you are located near a river and within 50 miles of a dam, you are exposed to an elevated risk of flooding. If the dam were breached, the water could be released and could cause an emergency. Check with your local planning agencies for their emergency action plans. If you are uncertain about the existence of dams, visit the Association of State Dam Safety Officials (ASDSO) at <http://www.damsafety.org> or the National Inventory of Dams at <http://crunch.tec.army.mil/nid/webpages/nid.cfm.>
>
> This web site is a database of over 76,000 dams in the country. Unfortunately, due to the potential for terrorist activities, the web site is no longer open to the general public. You may visit the site and leave a message for the Pagemaster. They will help you determine the state official who will be able to answer your questions. In the United States, thousands of dams exist and over half of them have Emergency Action Plans that identify areas downstream that would be flooded in the event of a breach.

Hurricanes

Facts and Figures

Hurricanes are one of the most devastating and costly of all natural weather events. This phenomenon thrives in the warm waters of the Atlantic and Caribbean Oceans as well as the Gulf of Mexico. In its early stages, this tropical disturbance is quite harmless. As it matures, however, an area of low pressure develops, winds begin a counter-clockwise rotary circulation, and wind speeds up to 35 mph are clocked. At this point the tropical disturbance is identified as a tropical depression. As the winds grow yet stronger (36 to 73 mph), the depression is categorized as a tropical storm. Once the wind velocity exceeds 73 mph, the storm graduates into an official named hurricane.

The hurricane season as defined by NOAA's National Hurricane Center lasts from June through November. During an average year, approximately ten tropical storms develop over the Atlantic Ocean and Gulf of Mexico. Typically six of these storms develop into hurricanes, and every three years about five hurricanes reach the United States. Two of these five hurricanes are usually considered major hurricanes with a Saffir-Simpson hurricane intensity scale rating of Category 3 or stronger.

Most emergencies or disasters happen without advance notice. The earth's tectonic plates suddenly shift and an earthquake results. A dam bursts and water floods a low-lying valley, surprising its inhabits. A tornado quickly develops and touches down while a community sleeps. Hurricanes, however, are unique because of their advance notice. We have the ability to track them as they develop. We can forecast their paths days in advance.

Saffir-Simpson Rating and Storm Categories

1. Minimal—Winds of 74 to 95 mph. Damage is minimal, usually limited to trees and power lines.

2. Moderate—Winds of 96 to 110 mph. Some roof damage may occur. Trees can be uprooted and power poles can be downed. Windows and storefronts can be damaged.

3. Extensive—Winds of 111 to 130 mph. Roofs are badly damaged or lost. Structural damage is common.

4. Extreme—Winds of 131 to 155 mph. Damage to most structures is severe. Extreme flooding can occur along and near the coast. Loss of life is common.

5. Catastrophic—Winds in excess of 155 mph. Damage is total in many places. Many structures are destroyed. Severe flooding is common several miles inland.

Logically, it would seem that, based on the advance notice and the ability to track a hurricane's path, very few lives would be lost and advance measures could be taken to mitigate property damage. In recent years we have learned much. The very fact that you are reading this book shows the importance placed on advanced planning for all emergencies. Building codes reflect the lessons we've learned from storms past; however, two factors keep hurricanes in the headlines and cause them to continue to take lives.

First, the population in the coastal flood plain areas has grown substantially. The homes and offices built in these locations are larger and more costly than years past. These prime slices of real estate don't house inexpensive metal buildings and dilapidated residential shacks. They are home to some of the finest and most extravagant residences and commercial sites imaginable. Unfortunately, when a storm of magnitude sweeps through these costal areas, it inflicts great damage to these properties. Because the buildings are bigger, the property has a higher value, and the replacement costs quickly soar.

The second factor that plays a role in the high costs of hurricane damage and loss of life is complacency. Residents of coastal areas become accustomed to numerous tropical storms and hurricanes that develop only to fizzle before they hit land. Those that do touch United States soil usually end up somewhere else, never in our own backyard. This fact is made evident each year as hurricane season begins and the first storm approaches. The day prior to the storm, people flock to lumber yards waiting hours for overpriced plywood for shuttering doors and windows. Many people are content to ride the storm out instead of heading inland or to the emergency shelters. Each time a near miss occurs, it simply reinforces the complacent attitudes.

Case Study 1: Hurricane Andrew

The classic example of the destructive force of a hurricane occurred on August 24, 1992, in Homestead, Florida. Hurricane Andrew caused an estimated $26 billion in damages in the United States alone, giving it the dubious distinction of being the most expensive natural disaster in U.S. history. Fortunately, Hurricane Andrew hit southern Dade County, a less populated and developed area, thereby lessening the economic impact of the disaster. If Hurricane Andrew had touched down just fifty miles to the north, billions of dollars more damage would have certainly occurred.

Hurricane Andrew had maximum sustained winds of 145 mph and gusts of 175 mph. Additionally, Hurricane Andrew brought with it a 16.9 foot storm surge. Hurricane Andrew destroyed 25,524 homes and damaged an additional 101,241. Twenty-three people died directly as a result of Andrew and 38 others died from secondary causes. Andrew disrupted electrical service to 1.4 million customers; many of these customers went weeks without electrical service.

> As devastating as Hurricane Andrew was, it never achieved Category 5 status. Through 1998 only 22 Atlantic storms have reached Category 5 strength, and of these 22, only two made U.S. landfall: the 1935 Florida Keys hurricane (not named) and Hurricane Camille, which pummeled the Mississippi coastline in 1969.

Case Study 2: Tropical Storm Allison, 2001

Obviously, hurricanes such as Andrew possess the ability to impart incredible economic damage. Although not usually as destructive, tropical storms also have the ability to inflict an incredible amount of damage. In June 2001, Tropical Storm Allison caused nearly $4.8 million in damage to the Houston area, making it the most expensive tropical storm in U.S. history.

Nearly 13,000 homes were completely destroyed or sustained major damage. The tropical storm spawned tornadoes, lightning, and floods, which claimed at least 41 lives. The Port of Houston received 37 inches of rain during the storm. Houston Heights in Harris County was hit with nearly 20 inches of rain in a 24-hour period.

Understanding the Terminology

To properly prepare for storm season, you must be familiar with the weather-related terminology used. Not understanding the difference between a tropical storm watch and a hurricane warning can be deadly. Storm advisories are issued by the National Weather Service (NWS). The following terms are used to describe a storm's threat.

- Tropical Storm Watch
  Within the next 36 hours, tropical storm conditions (winds from 36 to 73 mph) are possible in the storm watch area.
- Tropical Storm Warning
  Within the next 24 hours, tropical storm conditions (winds from 36 to 73 mph) are expected in the storm warning area.
- Hurricane Watch
  Within the next 36 hours, hurricane conditions (sustained winds greater then 73 mph) are possible in the hurricane watch area.
- Hurricane Warning
  Within the next 24 hours, hurricane conditions (sustained winds greater than 73 mph) are expected in the hurricane warning area.

When we think of hurricanes we usually think of high winds; however, hurricanes often produce torrential rains, flash floods, storm surges, and tornadoes. Although the strong winds of a hurricane cause incredible damage, often it is the byproducts of the hurricane that inflict the greatest damage.

Additionally, the following severe weather watches and warnings may be issued:

- Flood Watch
  Floodwaters are possible in the given area of the watch and during the time frame identified in the flood watch. Flood watches are usually issued for flooding that is expected to occur at least six hours after heavy rains have ended.
- Flood Warning
  Flooding is actually occurring or is imminent in the warning area.

> The areas most often affected by hurricanes are usually coastal towns and cities with high populations and dense development. However, inland areas can also be damaged. In 1989, hurricane Hugo, a Category 4 storm, made a direct hit on Charleston, South Carolina, but caused damage as far inland as Charlotte, North Carolina, over 200 miles inland. Windows were broken in downtown office buildings and air handler units were blown off rooftops.

- Flash Flood Watch
  Flash flooding is possible in the watch area. Flash flood watches are usually issued for flooding that is expected to occur within six hours after heavy rains have ended.
- Flash Flood Warning
  Flash flooding is actually occurring or is imminent in the warning area.
- Tornado Watch
  Tornadoes are possible in the given area of the watch.
- Tornado Warning
  A tornado has actually been sighted by a spotter or by radar and is occurring or is imminent in the warning area.

What to Do if Faced with a Hurricane

Before the Storm

Proper planning prevents pain and panic. That's a saying that is tossed around quite often by those responsible for emergency planning, and it's true! A few hours of planning prior to an emergency event will pay for itself many times over. Although we can't eliminate storms, there are a number of proactive things that can be done that will decrease the impact.

One of the first resources should be your local emergency management office or the Red Cross. They can inform you of approved evacuation routes and established shelters. Additionally, they can assist in creating an actual emergency plan and conducting dry runs. Another great resource is the Federal Emergency Management Agency (FEMA). This organization is at the forefront of emergency planning and recovery and can be an invaluable source of information.

As mentioned, your local emergency management agency will assist you in finding nearby shelters. Understand that not every "well-built" building constitutes a safe haven from a storm. Large gymnasiums, metal buildings, and other long span buildings provide little protection from a storm of significant magnitude. Additionally, structures built without proper hurricane strapping are susceptible to wind lift. A reinforced concrete structure might seem like a great storm shelter; however, if the roof isn't properly attached to the walls, there's a good chance it may take flight during the storm. This was a common problem in Dade County during Hurricane Andrew. Many structures were still intact, but the roofs came off and became very dangerous missiles. Because of this problem, numerous revisions to the South Florida Building

> Evacuation routes should take you a minimum of fifty miles inland, and depending on the storm, that may not be enough. Hurricanes tend to weaken as they pass over land, but a large hurricane can be 600 miles in diameter.

34   Emergency Preparedness for Facilities

Code were made, and these standards became a model for many at-risk communities across the county.

In addition to identifying safe shelters, your local emergency management agency will also outline the proper evacuation routes to be used in the event of a storm. Sometimes evacuations are optional, while other times they are mandatory. Either way, you'll need to know the best and safest route out of town, especially if you are in a low-lying coastal area and a storm is imminent. The problem begins when you (and everybody else) wait until the very last moment to evacuate. Imagine thousands of cars clogging the one or two evacuation routes hours prior to a storm making landfall. Not a pretty site, but it happens.

Storm Activities

Immediately prior to, during, and after any emergency it is often difficult to think clearly and rationally. The time before a hurricane strikes affords you the ability to use a checklist to ensure all critical issues are dealt with. Don't leave storm preparation and recovery steps to chance! The form depicted in Figure 2-4 is a representation of a storm activity list which specifically addresses issues that must be addressed prior to the storm's arrival. (A full view of the form is included in the Appendix and on the CD-ROM.) The

Figure 2-4. Storm Activities List: Before the Storm

beauty of this list is its format. It identifies critical tasks that must be completed at the following time increments:

- 72 hours prior to the storm's arrival
- 36 hours prior to the storm's arrival
- 24 hours prior to the storm's arrival
- After the storm has passed

The activity list also identifies the responsible employee or contractor and includes space for their phone numbers. As each activity is completed, simply highlight (electronically or manually) and move on to the next activity. This activity list should be distributed to all of your emergency management team. This form works well as a discussion base during conference calls before and after the storm.

Figure 2-5 depicts the same storm activity list, except this version is to be used after the storm has passed. Again, as each of the activities is completed, simply highlight and move on to the next task.

Storm Activities	After	Responsible Employee	Contractor Company	Office Phone	Cell Phone
Grounds					
Shrubs and hedges	Survey site. Indicate all damage on site map. Report to landscape contractor				
Trees	Survey site. Indicate all damage on site map. Report to landscape contractor				
Light Poles & Flag Poles	Survey site. Indicate all damage on site map. Report to electrical contractor				
Newspaper stands	Return to original location				
Drainage	Unclog all drains				
Parking	Remove all debris. Indicate helicopter landing areas				
Access	Check all gates for proper operation				
Vehicle Fuel Tanks	Order fuel, if needed				
Building Exterior					
Shutters	Open/Remove all shutters. Store removable shutters in designated area. Open windows to air				
Window Leaks	Report all damage to Window contractor				
Rain Gutters	Remove all debris from gutter				
Sand Bags	Remove sand bags after water has subsided				
Water Pumps	Set-up and run until water subsides				
Building Interior					
Water System	Repair all damages to water system. Disperse water bottles. Replenish water bottles until water system				
Sewer System	Schedule pickups of Port-o-Potties until sewer system is operational				
A/C System	Repair all damage. (Check roof top units visually)				
Generator System	If activated, monitor operations. Check usage and refuel if necessary				
UPS System	Report all irregularities to UPS maintenance contractor				
Telephone System	Inform telecommunications contractor of any irregularities.				
Radio	Radio operators check in per schedule or as needed to report any concerns.				
Garbage Pickup/Removal	Call for pickups as needed during restoration period.				
Food Operations	Feed employees per schedule for duration of restoration process. Restock supplies as needed.				
Refrigerator	If no electricity, open only when needed. When electricity restored, set controls to normal position				
TV and Cable Systems	Repair any damage				
Fire Alarm System	Repair all damage. Replace locks on shut-off valves. Notify monitoring contractor when repairs are complete.				
Security System	Guards monitor buildings and report any irregularities.				
Storm Supply Storage	Facilitate smooth transition of storm material from staging area to area where needed				
Miscellaneous					
Critical Documents	Bring back once building is dry and secure				
Insurance	Take video of interior and exterior				
Inventory List	Replenish used stock after hurricane.				
Contractor Avaliability Li	Call contractors to assign tasks				

Figure 2-5. Storm Activities List: After the Storm

Contractor Availability

Obviously, your company will have key people assigned to lend their assistance during storm mode. Business units such as human resources, procurement, engineering, and facility management will play an important role in preparation and recovery. No doubt outside assistance will be needed as well. With this in mind, a list of available contractors should be created. The Contractor Availability Form depicted in Figure 2-6 works well for this task. Seventy-two hours prior to the storm's predicted landfall, contractors should be contacted and their availability pre- and post-storm confirmed. This critical step must be completed prior to the storm. Understand that contractors may agree to offer assistance after the storm, but because of circumstances out of their control, they may not be available. Access roads could be blocked, their offices could be heavily damaged, or some other problem might exist. For this reason, you need to have alternate contractors available.

Contractor Availability Form: First Alternate

Issue	Contractor	Contact	Office Phone	Cell Phone	AVAILABLE	NOT AVAILABLE	LEFT MESSAGE	Comments
Generator								
Landscaper								
Sign Maker								
Parking Lot Sweeper								
Barricade Supplier								
Tool Rental								
Security Guard Service								
Motorized Gate								
Fence Contractor								
Roofing Contractor								
Ice Machine								
Ice Supplier								
Dumpster Rental								
Refuse Company								
Trash Hauler								
Port-O-Pottie Supplier								
Computer Systems								
Computer Network Tech								
Phone Technician								
HVAC Contractor								
Electrical Contractor								
Engineering								
Architect								
Plumber								
Locksmith								
Glass Company								
Movers								
Carpet Cleaner								
Janitorial								
Laundry Services								
Elevator Service								
Fire Protection Vendor								
Security System								
Electric Utility								
Gas Utility								
Water Utility								
Phone Company								

Figure 2-6. Contractor Availability Form

Inventory of Materials

Recovery from a storm occurs more quickly if you have an inventory of equipment and tools to work with. Again, don't leave the creation of this inventory up to chance. In addition to the typical emergency inventory, employees involved in storm recovery should be responsible for their own "survival" package. Necessary items include

- Two pairs of sturdy shoes
- Multiple changes of clothes
- Toiletries
- Sunblock and hat
- Essential medicines (especially prescription)
- Replacement glasses or extra contact lenses
- Cell phone including extra batteries and car charger
- Automobile with fuel topped off
- Food and drink for the first day

> While conducting audits of your building and grounds, be aware of downed power lines and broken water or sewer lines. Report these problems immediately to the proper utility company, police, or fire department. Don't try to negotiate around downed electrical lines. Be safe—and smart!

During the Storm

You will be limited in what you can do during the storm. If you followed your plan, now is the time to ride out the storm in a safe shelter or an area out of the storm path. Now is not the time to let your guard down when it comes to safety. Be especially cautious when using candles or kerosene lamps. Many buildings (and people) have survived a storm but died in a fire resulting from carelessness with lamps or candles.

It is advisable to listen to the radio for updates during the storm. There is a good chance that radio transmissions will be interrupted, so don't be surprised if this option is not available. Radios capable of receiving NOAA weather transmissions are the greatest source of information.

If electrical power is lost, turn off major appliances at the breaker panel to avoid a power surge. Unplug sensitive equipment such as televisions and computers. If possible, move these items to upper level floors and away from windows and doors. As a side note, make sure data from your computer is properly backed up and perhaps even stored off-site.

Windows, skylights, doors, and other building openings often represent the weakest link in a building; therefore, it would be prudent to avoid such areas during a storm. Instead, take refuge in a closet or bathroom near the center of a building and away for exterior openings such as windows and doors.

After the Storm

At some point the storm will have passed and efforts to begin restoration should be started. The best source of information concerning the storm status is the NOAA radio or local radio programming. When the storm is passing through, beware of misinterpreting the eye of the storm as the end of the storm. Remember, a hurricane is produced when winds begin a counter-clockwise rotary circulation around a center point. Within this center point or eye, hurricane strength winds are absent, and as this eye passes over, you may be fooled into thinking the storm has passed. Again, don't go outside until the National Weather Service has given the all-clear.

When the storm has passed, the first responsibility is to assist those injured. The first hours immediately following a hurricane are usually chaotic. Many buildings will be without utilities including water, telephone, and electricity. Having a plan in place will help you to remain calm and focused on the most important tasks. The Post-storm Activity List (Figure 2.5, p. 35) is an example of an audit that will help to focus your recovery efforts.

> For areas susceptible to storms, keeping a stockpile of food and other necessities is essential. If you keep a supply of necessities, don't forget that most of these items have a limited shelf life. Plan to rotate items such as food and batteries on an annual basis.

If you were evacuated prior to the storm, you may not be able to immediately return to your home or place of business. In these affected areas access is often restricted by local police and National Guard. Keep this in mind if you leave your property. Will you be able to conduct business away from the office? Have you thought about a remote location for conducting business and staging?

When you do finally return to your home or business, beware of snakes and other animals that may have been driven to higher ground because of flooding. Bugs and mosquitoes will be attracted to the wet conditions, so make sure you have packed adequate protection. Also be sure to pack adequate sunscreen. You'll be spending more time outdoors while trying to evaluate the damage and make repairs. Many natural shade providers such as trees and brush will be ripped out and displaced by the velocity of the winds. The sun will do more than give you a nasty burn. Working in the sun will quickly dehydrate the body as well. Take care to listen to your body, hydrate when necessary, and take breaks whenever you feel the need. Protect yourself from the elements early on, and you'll be able to successfully move from the chaotic post-storm phase to the controlled and productive recovery stage.

After a hurricane, water seems to be everywhere. Whether from a storm surge or relentless rains, there is a good chance you will be surrounded by standing water. One of the best steps to solve this problem is to begin pumping water from the premises and into other areas such as retention ponds or off-site locations. Don't forget to unclog roof drains, gutters, and catch basins that might be impeding the flow of water from your surrounding area as well. If water has made its way into your building, and if it is not currently raining, open the windows to ventilate and dry the building.

Tornadoes

Facts and Figures

A tornado is a naturally occurring event identified by a rapidly rotating column of air starting from a thunderstorm in the clouds and extending down to the earth's surface. The most dangerous tornadoes can have winds speeds of over 250 mph. The path of destruction found in the wake of a tornado can be miles wide and in excess of 50 miles long. Because of the destructive nature of tornadoes, millions of dollars each year are lost to its effects, and on average 42 people die from tornadoes each year. In the United States, tornadoes are more frequent in the Southeast and Midwest but can occur anywhere.

Like hurricanes, tornadoes have a season. That season runs from March through August, but tornadoes could occur anytime if the weather conditions are just right.

Understanding the Terminology

Tornado Watch

Tornadoes are possible in the given area of the watch.

Tornado Warning

A tornado has actually been sighted by a spotter or by radar and is occurring or is imminent in the warning area.

Case Study 1: Super Outbreak, 1974

The Super Outbreak of 1974 was aptly named. Within a 16-hour period 148 tornados touched down in 13 states. At one point 15 tornados had touched down at the same time. The path of destruction covered more than 2,500 miles and irreversibly changed the lives of thousands of people.

Xenia, Ohio, was the location of the deadliest of all the twisters. This particular tornado was particularly violent and reached an F-5 rating on the Fujita scale (see Figure 2-7). Winds reached speeds of upwards of 318 mph. On Thursday, April 4, when the winds had calmed, people emerged from their homes in a state of shock. Large tanker trucks had been tossed about like leaves. Large trees had been uprooted and tossed thousands of feet away. Roofs were torn off of buildings. Hundreds of buildings were leveled. In just twenty or thirty seconds, the tornado took the lives of 33 people and leveled half the city. More than 1,300 buildings were destroyed including four schools.

Xenia was not the only town to experience a deadly and rare F-5 tornado. During the Super Outbreak a total of six F-5 tornados developed: two in Ohio, two in Alabama, one in Kentucky, and one in Indiana. The 148 tornados that made up the Super Outbreak combined to take 315 lives and injure 5,484 people. Damages totaled $600 million, making April 4, 1974, one of the deadliest and most expensive days of tornadoes ever.

Category F0	Gale tornado 40–72 mph	Light damage. Some sign damage. Tree branches broken and shallow-rooted trees will be toppled.
Category F1	Moderate tornado 73–112 mph	Moderate damage. Roofing tiles loosen and take flight. Moving automobiles pushed off roads. Mobil homes may come off foundations.
Category F2	Significant Tornado 113–157 mph	Significant damage. Large trees are uprooted. Mobil homes are demolished. Roofs are torn off of some homes.
Category F3	Severe Tornado 158–206 mph	Severe damage. Cars can be lifted into the air and thrown about. Small missiles are generated. Roofs on well-constructed buildings are torn off.
Category F4	Devastating Tornado 207–260 mph	Devastating damage. Large missiles are generated. Well-constructed buildings are demolished.
Category F5	Incredible Tornado 261–318 mph	Incredible damage. Trees will be debarked. Well-constructed buildings will be lifted off their foundations and travel until they break apart. Automobile-sized missiles will travel in excess of 100 yards.

Figure 2-7. Fujita Tornado Categories

Case Study 2: Oklahoma/Kansas, 1999

On Monday evening, May 3, 1999, a devastating tornado developed and quickly gained strength. In fact, the tornado clocked the highest ever-recorded wind speed—318 mph. This distinction earned it a F-5 rating on the Fujita scale. Interestingly, if the tornado winds were just one mph faster, the tornado would have been classified an F-6, the highest rating on the Fujita scale and previously thought to be an unreachable milestone.

Oklahoma City was hit particularly hard by this wave of violence, enduring 52 tornadoes in a very short period of time. The largest of these tornadoes stayed on the ground for nearly four hours as it cut a path of destruction across the infamous "Tornado Alley."

By the time the estimated 76 tornadoes had ended, more than 55 deaths and 675 injuries were recorded. Over 8,000 buildings were damaged or destroyed, which contributed to over $1.2 billion in damages.

What to Do if Faced with a Tornado

Because of the "surprise" element of tornadoes, advance preparation is a must. Basic to this preparation is an understanding of exactly where designated shelters are located. This includes not just at the workplace but at home as well. If you do not have a designated shelter nearby or the close proximity of a tornado precludes you from accessing the shelter, you must be aware of areas of safe refuge available within the building. Here are a few guidelines:

- If you spot a tornado, immediately call 911 to report it
- Stay away from windows and doors
- If available, move to the basement
- If a basement is not available, go to a small interior room
- Stay on the lowest level of the building
- Stay tuned to the television or radio for further instructions
- Have access to a first aid kit and stored food and water

A good source of tornado information can be found on the FEMA website at <www.fema.gov>. This site offers a publication that shows how you can protect yourself from tornadoes while at home. Within the pages of this publication, you will find a Wind Zone Map, which graphically shows the frequency and strength of storms across the country. This map includes over 40 years of tornado history and 100 years of hurricane history. An-

> Tornado Alley is commonly used to describe the states from Texas north to the Dakotas. The region's geographic makeup and climate is conducive to the production of most of the nation's tornadoes. Dry air from the Rockies to the west collides with warm air from the Gulf of Mexico, and the union is anything but pleasant. These conditions create violent thunderstorms that spawn the tornadoes that are so prevalent in the region.

other great source of historical information can be found at <www.tornadoproject.com>. This site allows you to drill down to a county level and lists all occurrences and the magnitudes of hurricanes since 1950.

Tsunamis

Facts and Figures

Tsunamis are large waves that are the result of a disturbance in the water. They are often incorrectly referred to as tidal waves. Tsunamis have no association with the tides or any weather phenomenon. A tsunami usually occurs as a result of an underwater earthquake; however, the violent agitation can also be caused volcanic eruptions and even large landslides. Much like a rock dropped in a pond which sends ripples in all directions, in this case, the ripples are caused by a very large disturbance. The waves of a tsunami can travel at speeds over 500 mph and reach heights of 100 feet! For obvious reasons, low-lying coastal planes are at greatest risk, particularly land situated less than 50 feet above sea level. The nature of a tsunami is not to pound the shore with a single wave. Instead, a series of waves will hit land at a frequency between five minutes and 90 minutes. In open water the waves of a tsunami may appear to be normal and not of great concern; however, as they approach the shallower water near the shore, they quickly grow in size and strength.

Tsunamis could occur in any of the major oceans of the world; however, they are most prevalent in the Pacific Ocean. The Pacific Ocean is home to more active volcanoes than any other location on the planet; for this reason the area is aptly named the "ring of fire." Coastal areas that border the Pacific Ocean and the many islands found within the ring of fire are particularly susceptible to the threat of a tsunami. Although rare, tsunamis could occur in the Atlantic oceans caused by earthquakes, landslides, and volcanic disruptions.

Because of the large waves that are produced as a result of a tsunami, the majority of the deaths are caused by drowning. Other injuries and deaths are the result of flooding, debris, downed powerlines, broken gas lines, and polluted water supplies. For these reasons, care must be taken after the tsunami to avoid danger.

One byproduct of an earthquake, especially in areas with hilly terrain, is a landslide. In a coastal area, soil could become unstable and slide into the ocean. If the landslide is large, the resulting disruption in the ocean could very easily result in a tsunami. In a similar scenario, flank failure of a volcano could also produce a tsunami.

Case Study 1: Alaskan Tsunami, 1964

On March 28, 1964, an earthquake occurred at Prince William Sound in Alaska which generated a tsunami with waves between ten and 20 feet high along parts of the California, Oregon, and Washington coasts. A total of 123 lives were lost and over $108 million in damage occurred. Alaska was the hardest hit with over $84 million in damage and a total of 106 deaths. Five and one-half hours after the Alaskan earthquake, a regional tsunami crashed into Hilo, Hawaii causing additional damage and death. In addition to the regional tsunami, local Alaskan tsunamis were triggered by landslides that occurred as a result of the Prince William Sound earthquake.

Case Study 2: Hilo, Hawaii, 1946

On April 1, 1946, a tsunami with waves of 20 to 32 feet crashed into Hilo, Hawaii, flooding the downtown area and killing more than159 people. The tsunami was a result of an earthquake in the Aleutian islands of Alaska. The earthquake measured 7.1 on the Richter scale, and within five hours the tsunami traveled from Alaska to the Hawaiian Islands. This proved to be the most deadly and destructive tsunami in recorded history. Water reached more than one-half mile inland and property damage reached over $26 million. One of the telltale signs preceding a tsunami is the receding water at a beach. This curious phenomenon often leaves reefs exposed and onlookers will walk out to the reefs. This happened in Hilo, and many people, especially school children, were killed by the approaching tsunami as they stood on the exposed reefs. Hilo, Hawaii, is now home to the Tsunami museum, a memorial to all those who died in this tragic event.

On May 23, 1960, another tsunami destroyed much of downtown Hilo. This century there have been over a dozen significant tsunamis that have impacted the vulnerable Hawaiian Islands. Because of this, an early warning system has been implemented.

What to Do if faced with a Tsunami

Proper planning is critical for all emergencies, and tsunamis are no exception.

The logical starting point for tsunami preparation is to first determine if you are in an area that is susceptible to tsunamis. A good web site that provides historical information is the Russian Academy of Sciences' "Tsunami Laboratory" (<http://omzg.sscc.ru/tsulab/recent.html>). If tsunamis have occurred in your area, local communities will have a warning system in place. Take time to learn about these early warning systems.

In both case studies above, damage and death was high because of a lack of an early warning system. In 1965 a formal agreement was made between a number of tsunami tracking agencies including the Alaskan tsunami warning center and the Hawaiian tsunami warning center. Today, both centers are operated through the National Oceanic and Atmospheric Administration's (NOAA) National Weather Service. The Alaskan center serves as an early warning resource for Alaska, British Columbia, California, Oregon, and Washington State. The Hawaiian center serves as an early warning center for Hawaii and other Pacific-specific areas. The International Tsunami Information Center (ITIC) provides tsunami preparedness information to all of the Pacific Ocean nations.

The creation of these three agencies has been a life saver. Since the Pacific Tsunami Warning Center was established in 1948, 20 tsunami warnings have been issued. A tsunami has not been triggered without the Pacific Center knowing about it and issuing an advance warning; however, not everybody has heeded the warning and lives have been lost.

A Tsunami Watch is automatically declared when an earthquake occurs which meets certain criteria. The earthquake must be in an area that is likely to trigger a tsunami, and it must have a magnitude of 7.5 or greater on the Richter scale. For those earthquakes originating in the Aleutian Islands of Alaska, an earthquake 7.0 or greater will automatically trigger a Tsunami Watch.

At this point, notification of various agencies begins, and a limited public announcement is made.

> The three early warning centers can be reached at:
>
> West Coast Tsunami Warning Center of NOAA (includes Alaska)
> 910 S. Felton Street
> Palmer AK 99645
> www.tsunami.gov
>
> Pacific Tsunami Warning Center (Hawaii)
> 91-270 Fort Weaver Rd.
> Ewa Beach, HI 96706
> http://www.prh.noaa.gov/pr/ptwc/
>
> International Tsunami Information Center (Pacific-wide)
> 737 Bishop St. Suite 2200
> Honolulu, HI 96813-3213
> www.prh.noaa.gov/pr/itic/index.htm

The progress of the tsunami is monitored and readings from various tidal gauges are examined. If a tsunami can be confirmed, a Tsunami Warning is given. If the tsunami cannot be confirmed, the Tsunami Watch is cancelled. Because of the quick-moving nature of the tsunami, a Tsunami Warning warrants quick action. The public will be warned via the emergency broadcast system and evacuation will be implemented.

The next logical step in planning for possible tsunamis is to map out an elevated evacuation route. Because of the size of a tsunami, you should move to an elevation at least 50 feet above sea level and one mile inland. Don't limit yourself to a single evacuation path because that road may be blocked or unavailable during an evacuation. In Hawaii, predetermined evacuation routes have been established using historical and predictive data. The routes are easily identified and can usually be found in map form in the local telephone book.

When a tsunami occurs in open water, it can be tracked and early warnings can be given to those in its path. The United States has two early warning centers, one in Hawaii and one in Alaska. The NOAA weather radio is a great source for these warnings. If you are in an area susceptible to tsunamis or if you experience an earthquake, be ready to tune into your weather radio and evacuate to higher ground if instructed.

If you are in an area that could face a tsunami, it would be wise to have an emergency survival kit on hand. This survival kit might include
- Non-perishable food
- Nonelectric can opener
- Water

- Portable battery-powered radio
- Flashlight
- Extra batteries
- Basic tools for utility shut-off
- Personal medicine
- General first aid kit
- Clothing, including sturdy shoes
- Money

After the tsunami has reached land and the danger has passed, the first responsibility is to assist those injured. Do not attempt to move those seriously injured unless their location places them in additional danger. The following items should be addressed as well:

- Check for gas and water leaks. Turn off or report to utility company immediately.
- Check for downed power lines. Report to utility company.
- Examine the building for obvious structural damage. Warn others.
- Tune into your radio for further information.
- Cooperate fully with public safety officials.
- If possible stay off the roads. Emergency personnel should be given first priority.

After any emergency event, employees, employers, and family members should contact one another. Any delay in doing so might result in unnecessary search and concern. Develop an emergency communication plan that will allow you to contact these people immediately before and after the emergency event.

Tsunamis have been known to attract surfers and other curious onlookers to the beach. Make no mistake; the beach is the wrong place to be during a tsunami. As tsunamis get closer to shore, they often double in size and quickly overtake the beach area. Do not attempt to return to the beach after the initial impact of the tsunami. The waves travel in groups, and often the secondary waves are greater than the first. An extended period of time may elapse between the waves. Return to your buildings only when advised to do so by authorities.

After the tsunami has passed and you are given permission to return to your buildings, immediately evaluate the building's condition. If any doubt exists as to the safety or structural integrity of the building, do not enter it until professional advice has been given. If the building is safe to enter, you should open all doors and windows to dry out the building. If water has saturated the carpets, have them cleaned and treated immediately. If water is still standing onsite, set up sump pumps to evacuate the water. If extensive water damage has occurred, aggressive cleanup steps must be taken followed by testing for indoor air quality. If the damp environment remains, mold and mildew could develop, creating unwanted health problems and even greater cleanup costs.

Volcanoes

Facts and Figures

A volcano is a mountain that has an opening above a large pool of molten lava deep beneath the earth's surface. In its dormant state, little or no danger exists; however, that can quickly change. An eruption occurs when pressure builds up and the opening to the earth's surface provides the quickest escape route for the built-up pressure. The magma begins to rise and gasses begin to spew forth. A violent eruption will cause large rocks to explode, thereby becoming missiles traveling dangerously through the air. As rocks begin to fracture and shift, earthquakes can occur. In many cases, flank failure happens as well. Flank failure occurs when a large portion of the volcanic mountain slips away from the rest of the mountain, much like a huge avalanche. If the volcano is situated near the ocean and debris from this failure lands in the ocean, a tsunami is likely to occur.

Although hot lava and exploding rocks pose the obvious danger, debris flow can be equally devastating. Debris flow occurs when water, ash, rock, trees, and other items begin to slide down the volcano side. As the debris gains momentum and grows larger, its takes over everything in its path. The debris flow can be deceiving. It may develop in an area totally unaffected by lava flow. This is when onlookers get hurt. Volcanoes are very serious, very dangerous, and should be respected.

Because of the makeup of the earth, the area of the United States most likely to experience volcanic activity includes Hawaii, Alaska, California, Oregon, and Washington.

This area, minus Hawaii, makes up half of the "ring of fire." The second half of the ring of fire extends across the Pacific Ocean to the Orient. Hawaii is situated in the center of the "ring of fire."

Case Study 1: Mount St. Helens, 1980

On May 18, 1980, Mount St. Helens in Washington State violently erupted after more than a century of dormancy. Mount St. Helens quickly gained recognition as the most destructive volcano in the history of the United States. It caused more than $1 billion in property damage and took the lives of 58 people.

Typical of most volcanoes, most deaths were not the result of burning lava but instead were the result of massive mud and debris flow. The enormous landslide filled the valley area below, moving some 13 miles down the Toutle River and covering 24 square miles.

By the next day the volcano was no longer erupting but it continued to spew volcanic ash. The ash quickly traveled over the state of Washington, inducing darkness in many areas. In fact, roosters were fooled into thinking it was morning as they crowed in the mid-afternoon. The ash continued to travel over the central and northern United States. Within two weeks, a small trace of volcanic ash had drifted completely around the earth.

Case Study 2: Mauna Loa, Hawaii

Mauna Loa is located on the Big Island of Hawaii and is recognized as the world's largest active volcano. Mauna Loa is an impressive 13,677 feet above sea level; however, since the Hawaiian Islands are actually the tops of volcanoes, you could make a case that Mauna Loa should be measured from its true base at the bottom of the ocean floor to its crest. If you were to follow that logic, Mauna Loa would measure nearly 30,000 feet tall, higher than Mount Everest! Mauna Loa and neighboring Kilauea are two of the world's most active volcanoes. A drive through Volcano National Park will take you through a number of old lava fields, and if the conditions are right, you'll witness active lava flow spilling into the ocean below. Nearly eight miles of the Chain of Craters Road has been covered over by the active lava flow of Kilauea volcano. Because of this active flow, Hawaii is actually growing in size each year.

What to Do if Faced with a Volcano

The logical starting point for volcano preparation is to first determine if you are in a danger zone. The United States Department of the Interior, U.S. Geological Survey provides historical volcano information at <http://volcanoes.usgs.gov/Volcanoes/Historical.html>. If volcanoes are present in your area, local communities will have a warning system in place. Take time to learn about these early warning systems.

The next logical step is to map out an evacuation route. Don't limit yourself to a single exit path, because that road may be blocked or unavailable during an eruption. Avoid low-lying areas and, if possible, travel to higher regions. This suggestion will keep you from land that could quickly be filled with water or debris during a flash flood, but also realize that travel to higher areas may place you closer to the danger zone.

During and after a volcanic eruption, a number of other related natural emergencies may occur. Earthquakes are a common byproduct of a volcano. Landslides, mudflows, flash floods, fires, and tsunamis may also occur after volcanic activity. For this reason, you should continue to listen to your radio and be prepared to enact other portions of your safety plan.

If you are in an area that could be affected by a volcano, it would be wise to have an emergency survival kit on hand. This survival kit might include

- Non-perishable food
- Manual can opener
- Water
- Goggles
- Portable battery-powered radio
- Flashlight
- Extra batteries

- Basic tools for utility shut-off
- Personal medicine
- General first aid kit
- Clothing, including sturdy shoes
- Money

It is difficult to predict the volcano's path and reach; therefore, you should listen to your radio for instructions from the authorities. They will provide guidance concerning safe evacuation routes. Since a volcano's danger area is considered to be a 20-mile radius, those in the danger zone will likely be instructed to evacuate. If you are instructed to evacuate, do so. Don't believe that your building will provide adequate shelter. A violent volcanic eruption will send large boulders flying through the air. If these rocks were to hit your building, the structure will not likely provide you with a safe haven. If you are unable to evacuate, or if you are instructed to remain in place, you should close all windows, dampers, chimney flues, and doors. Extra precaution should be taken by placing wet towels under doors and window openings.

Since a volcano releases toxic fumes and glass-containing particles, you should avoid areas downwind of the volcano. If you are in an area where the breathing of fumes cannot be avoided, protect yourself with a respirator. If a respirator is not readily available, a dust mask might be used as a temporary solution while evacuating. If you are outdoors during a volcanic eruption and you are unable to immediately seek shelter, you should fashion a breathing mask using a shirt or other material. If water is available, soak the shirt first. Beware of low-lying areas where flash floods could occur. If the volcanic explosion produces a hail of rocks and debris, immediately search for an area that may provide relief from the rock storm.

During many natural emergencies, the curious gather to witness the event. All emergencies, especially volcanoes, are deadly and should be avoided at all costs. The danger of heat, lava flow, debris flow, flying rocks, and toxic fumes is real. What appears to be a safe vantage point may, in fact, place you in harm's way.

After the initial eruption, danger is still present from the volcanic ash and poisonous gasses. Stay tuned to your radio for further instructions. Do not attempt to re-enter the danger zone until the all clear has been given by the authorities. Because the volcanic ash may continue to flow even after you are allowed to re-enter your location, you must take safety precautions. Always wear a respirator and have extra filter cartridges. Because our eyes are perhaps the most sensitive organs we have, be sure to wear safety goggles at all times. Hard-soled shoes, long pants, and long-sleeves shirts are also recommended for safety reasons. For those with allergies or asthma, it would be advisable to not enter the affected areas until the ash flow has ceased and the surrounding areas have been cleaned.

In addition to the danger to humans, volcanic ash can create problems with machinery. Ash can clog engines, especially when a vehicle stirs up fallen ash on the road. Volcanic ash should be immediately removed from gutters and roofs. If allowed to accumulate, the heavy volcanic ash could cause a roof collapse or even structural failure. Although the danger area around a volcano usually covers about a 20-mile radius, disruptive volcanic ash can travel hundreds of miles.

Summary

When discussing emergency events, those considered natural events are the most common, deadly, and costly. No region in the country is exempt from dangerous natural events, and for this reason you must be prepared. Preparation usually starts with an awareness of the potential events that are likely to affect your area. Familiarize yourself with the weather-related terminologies such as watches, warnings, and advisories. Understand the necessary actions that should be taken when each of these terms are delivered by the authorities. Many good lessons can be learned by examining history. A number of good websites exist that can provide good information on emergency planning and recovery. Don't hesitate to use these resources when developing your emergency action plan.

3

Non-Natural Events

When we think of emergencies, too often we think of weather-related events. Obviously, emergencies and disasters can take the form of bomb threat, terrorist activities, nuclear accidents, building collapse, and other non-weather-related events. These events usually involve some form of human interaction and therefore have some predictability. Unfortunately, we often don't read the signs that precede the emergency. A disgruntled employee makes violent comments, but we simply shrug them off. Terrorists plant a bomb in the bowels of the World Trade Center in a semi-successful operation, but come back nine years later to finish the job. Not all non-natural events are predictable, but many of them are. Many of these events can be mitigated. Contingency plans can be created. Recovery and continuity plans can be written. The goal of this section is to identify non-natural events and discuss the proactive steps we can take to eliminate or minimize their occurrences.

Bomb Threat

Facts and Figures

Thousands of bomb attacks occur each year in the United States. Many of these bombs are crude, cause minimal damage, and are instigated by amateur citizens. Many of these attacks are small, localized in nature, and target buildings rather than people. For these reasons, most of these events do not make the national news. If the United States experiences thousands of bomb attacks per year, we can surmise that tens of thousands of bomb threats occur annually.

Obviously, bomb threats that are carried out are very costly, both in human life and property damage. However, even the threat of a bomb can interrupt work, introduce fear into the workplace, and, if the employer is perceived to handle the threat carelessly, erode employee confidence. Based on recent experiences, the threat of bomb attack—particularly those attributed to international terrorists—gives cause for great concern. Let's examine three case studies of bomb attacks that were watershed events, each of which grew progressively more deadly.

Case Study 1: World Trade Center, 1993

In retrospect, the 1993 World Trade Center bombing served as an eerie precursor to the 2001 attack that brought the World Trade Center towers down. The World Trade Center served as an attractive target, immediately recognized the world over as a symbol of American capitalism and located in the world's greatest city—New York City. In the case of the 1993 World Trade Center bombing, the perpetrators were able to hide a bomb in a vehicle and park that vehicle two stories below the ground level in the parking garage of the World Trade Center. Six people lost their lives and over a thousand were injured, mostly from smoke inhalation.

This scenario could have been much worse if the building structure were not as solid as it was and if the bomb used were greater in size. The bomb used 1,000 pounds of nitrourea and a few cylinders of hydrogen gas. It was believed that the terrorists responsible for this attack had planned for a larger bomb but did not have the financial resources to make it a reality. In fact, the terrorists were caught because they attempted to get their rental deposit back from the van rental company. Many believe that this bombing, although limited in success, proved to various terrorists cells that the United States was not untouchable and that if adequate funds were available, the towers could be felled.

Case Study 2: Murrah Building, 1995

One of the best defenses against a bomb attack is distance. If you can keep distance between the bomb and the building (and personnel), you can limit the effects of the bomb. In 1995 a domestic terrorist was able to load a moving van full of explosives within 20 feet of the Alfred P. Murrah federal building in Oklahoma City. The violent explosion resulted in a significant building collapse and the death of 168 people, including many children in a daycare center. Glass was broken more than a mile away. If the Murrah building had greater standoff areas around the building (in other words, if vehicle access had been limited near the building), the effects of the bomb might not have been as severe.

Two interesting facts became apparent with this event. First, a domestic terrorist was the mastermind of this attack. Usually, we fear international terrorists and give little concern to the actions of our own disgruntled citizens. Secondly, this attack did not occur in a large metropolitan area such as New York or Los Angeles. The attack occurred in the less-populated heartland of the United States. Who would have thought that a federal building in Oklahoma City would become an attractive target for a terrorist, domestic or otherwise? Any preconceived notions we may have had about terrorism had been shattered.

Case Study 3: World Trade Center, 2001

September 11, 2001—a day that will "live in infamy." That phrase was used once before, on December 7, 1941, immediately after the Japanese attacked Pearl Harbor. The attacks of September 11, 2001, would eclipse even that infamous day.

At 8:45 A.M. American Airlines Flight 11 out of Boston was hijacked and flown into the World Trade Center North Tower. All 81 passengers and 11 crew members were killed. The explosion was tremendous and left a gaping hole in the tower, taking the lives of many more inside the tower. Just minutes later, at 9:03, a second hijacked airliner, United Airlines flight 175, was flown into the South Tower, killing all 56 passengers and nine crewmembers. Within a half hour, the FAA shut down all air traffic throughout the country. At 9:43 American Airlines flight 77 was hijacked and flown into the Pentagon, severely damaging the two outermost rings and immediately killing all 58 passengers and six crewmembers aboard. Within a half hour a large portion of the damaged area collapsed.

At 10:05 the South Tower of the World Trade Center experienced total collapse. At 10:10 the final hijacked flight, United Airlines flight 93, crashed in a field southeast of Pittsburgh, Pennsylvania. At 10:28 the World Trade Center's North Tower collapsed. At 5:20 p.m. the 47-story Building 7 of the World Trade Center collapsed as a result of damage sustained earlier when the North and South Towers collapsed. In the next few days and weeks, the grisly task of searching for bodies and cleanup continued. The final tally of the attacks included 2,888 confirmed deaths.

Days later, a new threat was introduced: Letters tainted with anthrax spores added to the death toll. If you include the missing and presumed dead from the air attacks, the number of deaths rises past 3,000.

Aside from the loss of human life, the economic impact of the September 2001 attacks has been phenomenal. It will take some time to accurately account for the impact of these events. It will take even longer to recover. From a real estate standpoint, tens of millions of square feet of prime office space was lost in a day. Organizations were required to move to other sites or go out of business altogether.

Never before has this country had to be concerned with terrorism on such a large scale. Nor has the world, for that matter. Obviously the rules have changed. Whether the threat be a suicide bomber or a hijacked airplane, an anthrax laden letter or a crude nuclear device hidden in a suitcase, domestic terrorist or foreign, we must be aware of the possibilities.

What to Do if Faced with a Bomb Threat

The best defense against a bomb attack is preparation. People who have a plan for the eventuality of a bomb threat are in the best position to survive the threat of attack. Even if your facility is ultra-secure, it remains impossible to eliminate bomb threats. Threats can be carried out by anyone with access to a telephone. Even if an actual attack never materializes, the threat can cause major disruptions and expense. Upon receiving a threat, one must be prepared for some sort of action. The usual first course of action is evaluation. You must be prepared to evaluate the threat and its legitimacy. This is not an exact science and is open to interpretation, but without some standard method of evaluation you will be ill prepared for the threat.

The Bureau of Alcohol, Tobacco, and Firearms (ATF) has prepared an excellent pamphlet titled "Bomb Threats and Physical Security Planning" which can be requested from the ATF in writing at:

Bureau of Alcohol, Tobacco, and Firearms
Arson and Explosives Programs Division
800 K Street, NW, Tech World Suite 710
Washington, DC 20001

Among the many good suggestions, you will find that they have created a Bomb Threat Checklist that serves as a template for bomb threats received via telephone (see Figure 3-1).

The Bomb Threat Checklist

This checklist gives the recipient of a bomb threat a list of questions to ask the caller. Some questions are obvious such as "When is the bomb going to explode?" and "Where is the bomb located?" while other questions seem less obvious. Questions such as "What is your name?" and "Where are you calling from?" might not be answered by the caller, but stranger things have happened! Even if all of the questions on this checklist are not answered, it keeps the caller on the phone line for an extended period of time, and that will assist the authorities in tracing the call. Another section of the checklist is designed so the recipient can circle a description of the caller's voice. This may help to identify the caller and may assist you in determining the legitimacy of the threat. For instance, if the caller sounds stressed and is crying, you might decide the threat is a real one. However, if the caller sounds young and is giggling while speaking, you may determine that call is a prank.

This bomb threat checklist should be copied onto a bright colored paper that stands out amidst the typical clutter in an office. In the event of a threat by telephone, the recipient can quickly find the checklist and take the appropriate notes.

Whether to Evacuate

Because of increased demands on local police departments and other related agencies, the decision to search for a bomb and the call for an evacuation is one that will be made

ATF BOMB THREAT CHECKLIST

Exact time of call _____

Exact words of caller _____
_____
_____
_____

QUESTIONS TO ASK

1. When is bomb going to explode? _____

2. Where is the bomb? _____

3. What does it look like? _____

4. What kind of bomb is it? _____

5. What will cause it to explode? _____

6. Did you place the bomb? _____

7. Why? _____

8. Where are you calling from? _____

9. What is your address? _____

10. What is your name? _____

CALLER'S VOICE (circle)

Calm	Disguised	Nasal	Angry	Broken
Stutter	Slow	Sincere	Lisp	Rapid
Giggling	Deep	Crying	Squeaky	Excited
Stressed	Accent	Loud	Slurred	Normal

If voice is familiar, whom did it sound like? _____

Were there any background noises? _____

Remarks: _____
_____

Person receiving call: _____

Telephone number call received at: _____

Date: _____

Report call immediately to: _____
(Refer to bomb incident plan)

Detach and place by each telephone. Duplicate as necessary.

Figure 3-1. Bomb Threat Checklist for Phone

by you, the building owner, or manager. In the event a bomb threat has been received, a decision to evacuate or remain will need to be made quickly. The basis for this decision will often depend on the perceived credibility of the caller. When faced with a bomb threat, you have three basic options:
1. Ignore the threat.
2. Evacuate immediately.
3. Search for the bomb and evacuate if necessary.

Obviously each of these options presents its own pros and cons. You must take time to develop procedures prior to a bomb threat occurring. When developing these procedures, consult with your local fire and police departments. They will be able to provide suggestions and possibly even training. In many cities across the country, your local police or fire department will not conduct a bomb search if you receive a threat. They will expect your organization to make the decision to evacuate. They will expect your organization to perform the actual bomb search. This may seem unfair or even dangerous but it underscores the need for a game plan when dealing with these issues. If you do happen to find a bomb or something unusual, the police or fire department will defuse and remove the bomb, but again, this is after you find it! By consulting with your local fire department and police department, you'll likely be able to predetermine their level of involvement in a search as well.

After you have received a bomb threat, beware of touching or disturbing any suspicious packages. Clear the area around the suspicious package and contact the authorities immediately. Keep all streets and sidewalks clear for emergency vehicles and have a member of your search team wait for the authorities to arrive and provide an escort to the location in question.

The Bomb Search

At some point, especially if the threat seems to have merit, the decision to conduct a search for the bomb will need to be made. As mentioned previously, the actual bomb search will likely be conducted by the building manager or owner. If this describes your situation, now is the time to identify processes and procedures and communicate them to the responsible parties. Here are a few steps that must be addressed prior to the actual search:

- Provide each employee with a bomb threat checklist.
- Train employees how to evacuate the building in a calm and orderly fashion.
- Determine if a bomb squad is available to assist you and how they will respond. Police and fire departments are a good starting place.
- Designate search team members (include someone who works in the department or floor being searched).

- Have a set of floor plans available for the search team and authorities.
- Practice efficient search techniques.
- Establish a command center outside of the building.
- Establish a procedure to report progress back to authorities.
- Establish a process to allow employees back into the building.

The building search should be conducted in an orderly and thorough fashion. If a building search is to be conducted, the following tools should be made available and used:

- Keys: Some rooms may be locked and require a unique key. This is usually not a problem during regular operating hours but could be a real problem after hours.
- Stepladder: Many of the areas to be searched are above eye level and cannot be examined properly without a small step ladder.
- Flashlight: Not every area will be well lit, so be sure to have a flashlight handy.
- Mirrors: Mirrors are critical for gaining visual access to certain areas. Check with your local law enforcement agency for suppliers of such mirrors.
- Floor plan: A floor plan can be instrumental in conducting a bomb search. Be sure to have highlighters on hand to mark areas on the floor plan after they are searched. If an actual sweep and search of the premises becomes necessary, it should be conducted in an efficient and safe manner. The aforementioned "Bomb Threats and Physical Security Planning" pamphlet prepared by the Bureau of Alcohol, Tobacco, and Firearms (ATF) includes an excellent procedure for sweeping and searching buildings (See Figures 3-2 through 3-6). This procedure entails five steps:

Step #1: Enter the room. Stop, look, and listen.

Step #2: Divide the room by height for search.

     Zone 1: Floor to waist

     Zone 2: Waist to chin

     Zone 3: Chin to ceiling

     Zone 4: Above false (drop) ceiling

Step #3: Search room by zone and assigned area. Overlap if necessary.

Step #4: Search the interior public areas.

Step #5: Search the outside areas.

#1 ROOM SEARCH-STOP, LISTEN

Figure 3-2. Search Procedures Step 1

4 FALSE CEILING
3 TO CEILING
2 WAIST TO CHIN
1 FLOOR TO WAIST

#2 DIVIDE ROOM BY HEIGHT FOR SEARCH

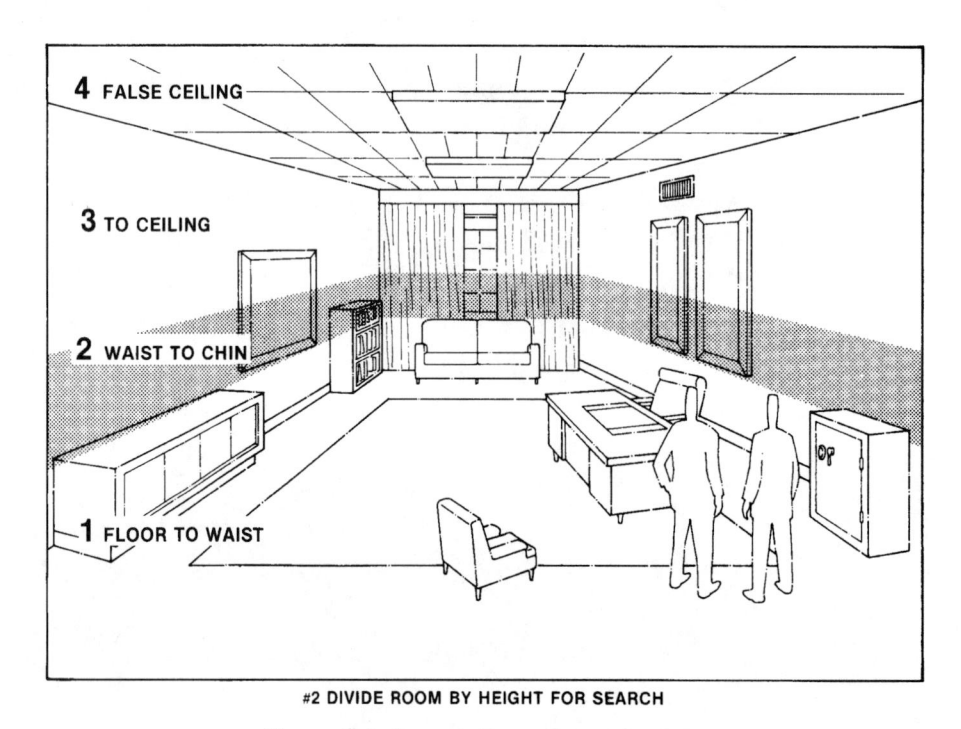

Figure 3-3. Search Procedures Step 2

#3 SEARCH ROOM BY HEIGHT & ASSIGNED AREA,
OVERLAP FOR BETTER COVERAGE

Figure 3-4. Search Procedures Step 3

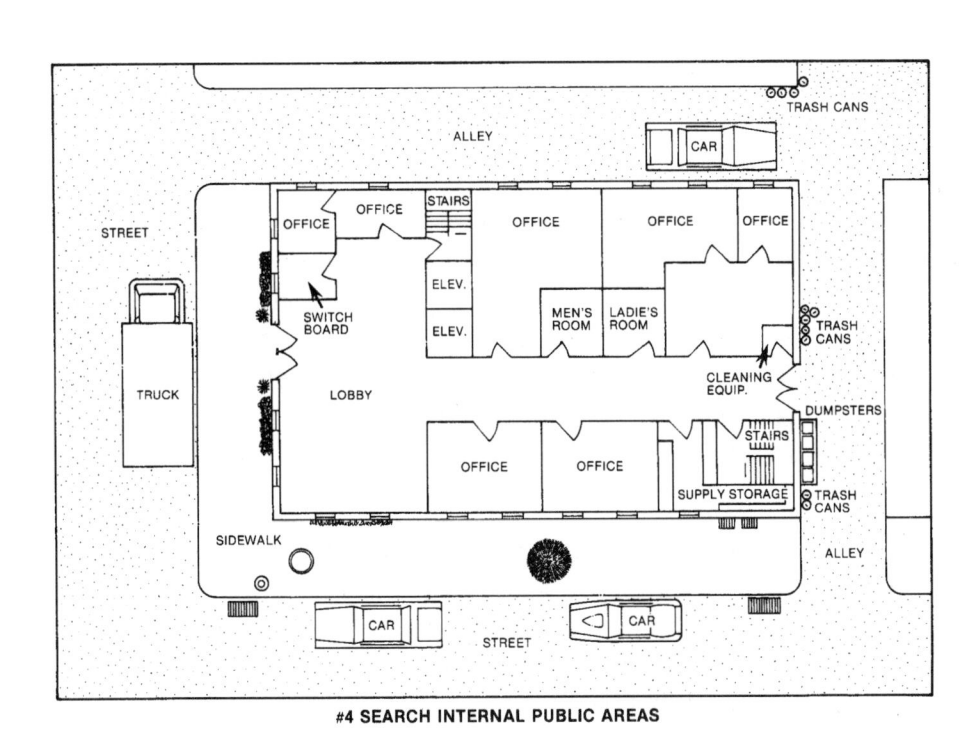

#4 SEARCH INTERNAL PUBLIC AREAS

Figure 3-5. Search Procedures Step 4

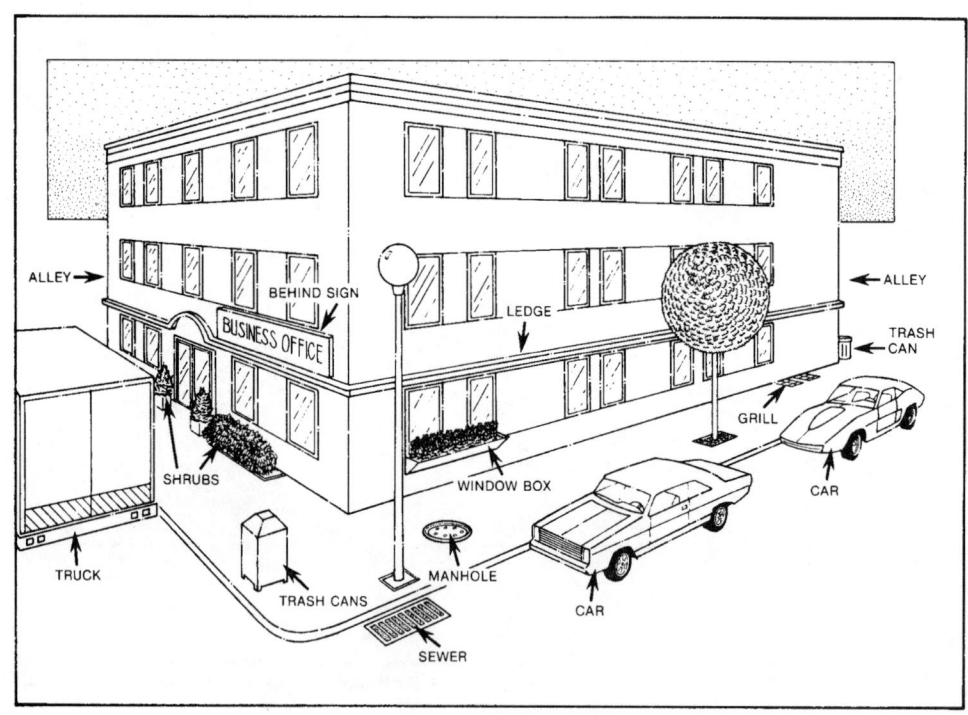

#5 SEARCH OUTSIDE AREAS

Figure 3-6. Search Procedures Step 5

Step number two is critical. A minimum of two people should enter a room and walk to one end of the room. The two employees will stand back to back and proceed around the perimeter of the room, limiting their observations to "Zone 1," which is from the floor to waist height. This might entail looking underneath desks, in garbage cans, behind plants, and other such areas. As the two employees complete their sweep of the perimeter of the room, they will eventually meet back at the starting point. This completes the wall sweep of Zone 1. Next, the employees would turn around, back to back again, and repeat the perimeter sweep once more. This time however, the two employees will observe all items in "Zone 2," which is waist height to chin height. Each step is repeated for each zone, and finally a sweep of the middle area of the room should be completed. This logical process should be completed in each room to ensure the absence of an explosive device.

When using a search team for the building sweep, it is recommended that you have an employee who works in that specific area on the team. That employee is most likely to know what belongs and what is out of place. When a specific department, floor, or common area has been completely searched, highlight that area on your floor plan and communicate back to central command that the area has been searched.

A command center must always be set up to help handle the evacuation, search, communication, and eventual re-entry into the building. The command center would include an employee with a floor plan stationed outside of the building structure. This person would keep track of the employees and visitors evacuated and the areas searched. They would serve as the liaison between the authorities and the search and sweep team. The command center would give the employees and visitors the "green light" to re-enter the building. For this reason, the command center must be staffed by a person(s) who has the authority to make "re-enter or go home" decisions. A definite chain of command must be established prior to attempting a sweep and search of evacuation. Without these critical issues being resolved, the process will be sloppy, chaotic, and will likely not be completed properly, thus placing employees at risk.

> A word of warning: When searching a building, beware of using two-way radios as a means of communication because these devices can cause a premature detonation of an electronic blasting cap! Instead of radios and cell phones, use a runner to communicate progress back to the command center.
>
> Upon establishing procedures for handling bomb threats and evacuations, share these procedures with your employees. They will be eager to hear your plan for dealing with these threats. An organization without bomb threat procedures stands the risk of losing the confidence of its employees. Worse still is the organization that has no plan and hides the occurrence of threats from its employees.
>
> If a bomb does explode, many employees will be disoriented and panicked. Instruct the search team to assist the employees in evacuating in an orderly fashion. Be prepared to isolate the damaged area and do not allow employees to enter the building until it has been inspected for structural integrity.

Chemical and Biological Concerns

Facts and Figures

If this book were written just a few years earlier, this section may not have been included because the threat of bioterrorism on American soil seemed so remote. Words such as anthrax and biotoxins were foreign to us just a short time ago. Today we enter a new reality, one in which we must be concerned about the use of biological organisms as weapons of mass destruction. Chemical and biological agents can be delivered a number of ways. From anthrax in an envelope to smallpox injected into another human, the threat of these concerns are very real. Making these threats even more alarming is the fact that many of these agents are invisible to the eye and don't manifest themselves for days or even weeks. We have vaccines that can effectively counteract or prevent illness for some of these agents, but for others we do not.

The intent of this section is not to scare the public nor is it to instill an air of hopelessness. The fact remains, widespread chemical and biological warfare is a possibility, although a remote one. Better to be armed with knowledge than to be caught unaware and unable to process the threat. Weigh both the liability and probability of such a threat, and if you deem it a possible threat to your business operations, incorporate these events into your emergency planning process.

Case Study 1: Anthrax Mail Threat, 2001

Never before in American history has the receipt of mail caused so much concern. In the 1990s, we were concerned over letter bomb packages sent by a mysterious person dubbed the "Unabomber." When the perpetrator was finally discovered and captured, we learned about a domestic terrorist named Ted Kaczynski. We were concerned, but Mr. Kaczynski seemed to be targeting specific people, namely scientists and professors. If that didn't describe us, we were not overly concerned. Additionally, we became educated about the dangers of unknown packages, particularly those large in size. It was not until the anthrax mail threats immediately following the terrorist activities of September 11, 2001, that we became really concerned about our mail delivery.

> The United States Postal Service provides a safety poster that identifies suspicious mail and how it should be handled. This poster is include on the CD-ROM accompanying this book and can also be downloaded at <www.usps.com/news/_pdf/poster.pdf>

What to Do if Faced with a Chemical or Biological Threat

The U.S. Postal Service has created posters to educate us about suspicious packages and how to handle them. One of the telltale signs of a potentially dangerous package is the use of excessive stamps. This may seem trivial, but most letters and packages sent for legitimate business purposes use metered mail and not stamps. The metered mail can be traced to its origin, something that stamps do not allow. In an effort to mail the piece, the sender will often put more stamps on the package than necessary. Other signs to look for include:

- Lack of return address
- Foreign country of origin
- Misspelled words
- Addressed to a title (i.e., CEO) instead of a personal name
- Warnings not to x-ray (i.e., film enclosed)
- Heavy or lopsided package

Many businesses have opted to open mail off site or to outsource mailroom operations altogether. If a chemical or biological threat is received by mail, exposure would be limited to the off-site location.

If you fear an exposure to chemical or biological agents, immediately seal or cover the envelope or package. Leave the room and if possible lock it to make sure no one else is able to come in contact with the package. Wash your hands with soap and water to prevent spreading. Notify your building manager and the local police department. Finally, make a list of everyone who was in the room when the package was opened. Pass that list to your local public health department.

The Center for Disease Control has created three categories for diseases and harmful agents. The categories and examples of each include the following:

> One of the greatest concerns for building managers is the potential for biological agents to be spread via the HVAC system. Many building owners and managers have created a process that will allow them to shut down the entire HVAC system (including intakes, dampers, etc.) with one flip of a switch. In the past, shut down required many switches be turned off, was time consuming, and was not always completed correctly.

Category A

Public healthcare providers must be prepared to address a number of biological agents, including pathogens that are rarely seen in the United States. High-priority agents pose a serious risk to the national security because they can be readily produced and easily disseminated from person to person, resulting in deaths and the potential for a major public health impact. Additionally, occurrence of these agents might cause public panic and social disruption.

Category A Examples:
- Anthrax
- Botulism
- Plague
- Smallpox
- Ebola

Category B

These agents comprise a second tier of priority scale. Typically they are moderately easy to disseminate and result in moderate morbidity rates.

Category B Examples:
- Food safety threats (Salmonella)
- Brucellosis
- Melioidosis
- Q fever
- Water safety threats

Category C

This third priority level includes agents and emerging pathogens that could be engineered for mass dissemination in the future because of their availability, production, and dissemination. With Category C diseases and agents, a potential for high morbidity exists and a major health impact could occur.

Category C Examples:

Emerging infectious disease threats such as Nipah virus and hantavirus

Because of recent terrorist related events, we have become somewhat familiar with a number of different chemical or biological concerns. The following is just a sampling of some of the potential dangers that exist.

> The Center For Disease Control recommends contacting your local county or city health departments if you believe you have been exposed to chemical or biological agents. A detailed listing of these departments can be viewed at the CDC website <http://www.bt.cdc.gov/EmContact/index.asp>. Additionally, the CDC emergency Response Hotline is available 24 hours a day at (770) 488-7100.

Anthrax

Bacterial spores often found in cows, sheep, and other grazing animals cause this deadly infection. Inhaled anthrax is usually fatal if not treated immediately. Treatment includes antibiotics such as ciprofloxacin, but it must be taken immediately. Symptoms begin as a common cold but quickly progress into breathing problems and shock. The spores are relatively easy to manufacture but difficult to disperse in a way that would cause death. Spores must be very fine to produce a powder-like substance which can float through the air and into nasal passages.

Anthrax can be destroyed by an irradiation device that focuses a beam of electrons on the target. A second way to kill anthrax is to use Sandia foam, a disinfectant that causes an oxidizing reaction. Finally, a third way to kill anthrax is the use of chlorine dioxide gas. This method is effective because the gas can be distributed in a building through the HVAC system.

Nerve Agents

In the 1930s, a German scientist named Gerhard Schrader was working toward developing chemical pesticides for crops. His efforts were overly successful; he produced the nerve agent Tabun. Mr. Schrader later produced the nerve agents Sarin and Soman.

A couple of noteworthy events involving nerve agents have occurred in recent times. During the Iran-Iraq war (1984–1988), Iraq used the nerve agent Tabun and a mixture of other agents against Iran and its people causing over 5,000 deaths and nearly 65,000 injuries. A few years later, the Aum Shinrikyo Cult was reported to have used the nerve agent Sarin in a Tokyo subway. Obviously, we have learned that small nations or organizations without much money or political clout can inflict tremendous harm upon their enemies.

Nerve agents such as Sarin, Tabim, and VX can affect all parts of the central nervous system. Symptoms include eye pain, blurred vision, tightness in chest, nausea, vomiting, tearing, muscle weakness, convulsions, loss of consciousness, and coma. Long-term effects include anxiety, insomnia, depression, headaches, slurred speech, and difficulty concentrating. Atropine is a common drug used to treat nerve agent exposure. It has proven successful in treating many of the symptoms of nerve agent exposure. Other treatments exist, but they have limitations.

Mustard Gas

In World War I, Germany favored the use of mustard gas because it was almost odorless and took twelve hours to take effect. Historians report that in World War II, Hitler was reluctant to use mustard gas because he had been victimized by chemical agents and was unwilling to introduce new and more toxic agents.

Although not normally deadly, mustard gas typically produces blisters and irritates the eyes and respiratory system. Large doses can cause burns and lung damage, can damage the immune system, and can be life threatening if the exposure is great. Unfortunately, there is no known antidote for mustard gas exposure. Decontamination must be completed within two minutes after exposure to limit damage. Immediately remove all clothing and begin the decontamination process. Decontamination includes washing with baking soda or Fuller's earth powder. If these products are unavailable, use water and soap. Be sure to irrigate eyes with saline solution, if available. If saline solution is not available, use clear water.

> For more comprehensive information on all types of chemical and biological information, visit the Center for Disease Control (CDC) and Prevention bioterrorism website at <www.bt.cdc.gov>. This is an excellent resource for bioterrorism preparedness and response and provides a host of downloadable files, videos, pictures, and FAQs.

Small Pox

This disease is universally feared as the most destructive of infectious diseases. It was eradicated from humans in the 1970s, although both the United States and Russia store the virus in laboratories. There is some concern over the security of such labs, especially those in Russia. It is thought that other countries may also have small samples of the smallpox virus.

There is no proven treatment for smallpox although vaccination within four days of exposure can lessen the effects. Within two weeks after a smallpox exposure, high fever and headache give way to more visible symptoms including a rash on the mouth and hands. The smallpox disease is fatal in about thirty percent of the exposures.

This is by no means an all-inclusive list of the chemical and biological threats to humans. Other threats such as cholera, which causes an acute diarrheal illness, are possible concerns but are not as deadly as other agents. A cholera exposure typically results in death in less than one percent of exposures when treated properly.

Chemical Spills and Contamination

Facts and Figures

A hazardous chemical spill can create dangers comparable to any other type of disaster, including the loss of life. From a small chemical spill on a factory floor that causes an evacuation of a few people to a large-scale release of dangerous chemicals that affects many people, you should be prepared for the potential of a spill. Not only is this smart business, it is also required by law.

A Philadelphia refinery stores 400,000 pounds of hydrogen fluorida that, if leaked, could asphyxiate close to four million nearby residents. The Atofina Chemicals, Inc., plant outside of Detroit has predicted that a single rupture of one of its 90-ton railcars of chlorine could affect and endanger three million people. According to the Environmental Protection Agency (EPA), at least 123 plants across the country keep amounts of toxic chemicals that, if released, could form deadly vapor clouds that would endanger over one million people. Additionally, at least 700 plants store chemicals that, if released, could endanger over 100,000 people.

The number of persons who could be injured following the release of chemicals and the extent of the damage will depend on numerous factors. Some factors are the properties of the substance, the size of the discharge, the period during which people are exposed, and the length of time that passes between exposure and treatment.

On December 23, 1984, workers at the Union Carbide factory in Bhopal, India, reported to their supervisors that their eyes were burning and tearing. The cause of the discomfort was a dangerous chemical reaction that occurred when a large amount of water got into a methyl isocyanate (MIC) storage tank. For nearly two hours, 40 tons of methyl isocyanate poured out of the tank and escaped into the air, creating a cloud of poisonous gas over the city of nearly one million people. Because the accident occurred in the middle of the night, many of the city's inhabitants died in their sleep. All told, estimated deaths reached upwards of 4,000 people, and hundreds of thousands of people were sickened. Many of the victims were people living in poor shantytowns surrounding the Union Carbide plant. Many of the people affected in 1984 are still feeling the effects of the poisonous gas even today.

A 1990 study commissioned by EPA found that since 1980 there were at least 15 accidents in the United States that exceeded the Bhopal, India, release in volume and toxicity of chemicals released. Only circumstances such as wind conditions, containment measures, rapid evacuations, and facility location prevented disastrous consequences from taking place.

A recent study by the U.S. Chemical Safety Board found that between 1987 and 1996, there were on average 60,000 industrial chemical accidents every year, and that more than 250 people died each year as a result of such accidents. These incidents include a range of events, not all of which necessarily resulted in consequences like injuries, deaths, or evacuations. More than 4,800 facilities in the United States each store at least 100,000 pounds of an extremely hazardous substance, which is more than the amount of volatile toxic chemicals released at Bhopal (some 90,000 pounds).

In 1990 as a part of the Clean Air Act Amendments Section 112(r), the U.S. Congress required industrial sites that use extremely hazardous substances to disclose Risk Management Plans (RMPs). The RMPs have three parts: First, a hazard assessment describes release scenarios, potential off-site consequences, and a five-year accident history; second, a prevention program addresses basic safety procedures such as training, maintenance, and safety audits; and third, an emergency response program covers response plans, drills, and coordination with local planners.

Industry representatives oppose a national public data system that includes placing all parts of the RMP on the Internet. Chemical industry lobbyists argue that publicizing worst-case scenarios will lead to terrorist attacks targeted at their facilities. Their solution to ensuring public safety is keeping this information off the Internet. In August 1999, Congress restricted public access to these plans. However, the executive summaries for the RMPs by state including worst-case scenarios can be viewed at the Right to Know Network website. (Available at <http://db.rtknet.org/T1050>.)

A number of federal laws regulate hazardous materials including the Occupational Safety and Health Act (OSH Act) Subpart H (29 CFR 1910.101–1910.126). Specifically, 1910.119 Appendix A provides a list of toxic and reactive chemicals that present a potential for a catastrophic event.

Other Federal laws include the Superfund Amendments and Reauthorization Act of 1986 (SARA) and the Resource Conservation and Recovery Act of 1976 (RCRA). Like OSH Act, Title III of SARA regulates the packaging, handling, labeling, transportation, and storage of hazardous chemicals.

Case Study 1: The Love Canal, 1976

The infamous Love Canal Treatment Facility in Niagara Falls, New York, is perhaps the best (or worst) example of man's lack of concern for the earth's environment. During the 1940s and early 1950s, Hooker Chemicals used the landfill extensively. When the site reached its maximum capacity, Hooker Chemical covered the landfill with dirt. Soon thereafter they sold the site to the Niagara Falls Board of Education for one dollar.

> Material Safety Data Sheets are available for just about every chemical produced. Upon accepting delivery of a chemical, make sure the MSDS is included. When purchasing small quantities of a chemical from a retail store, ask a clerk for a copy. If you have existing chemicals on site without an MSDS, visit <www.msdssearch.com/>, provide some basic information, and you'll be able to print out the needed MSDS.

Eventually a public school was built on the site in addition to hundreds of single-family homes. As early as 1976, residents began to question the unknown substances seeping into their basements and storm sewers. Eventually, the school was closed and over 700 families were relocated.

Nearly 21,000 tons of toxic chemicals were dumped at the 15-acre Love Canal landfill. Over 200 different chemical compounds are identified as being buried there. In August 1978, President Carter declared a federal emergency at the Love Canal. In 1979 a federal report estimated the odds of Love Canal residents contracting cancer to be as high as 1 in 10. In 1980 the EPA discovered that chromosome damage was found in 11 out of 36 residents tested from the Love Canal area.

Case Study 2: General Chemical, 1993

On July 26, 1993, a leak of oleum (concentrated sulfuric acid and sulfuric trioxide) from a railroad tank car at a manufacturing plant operated by the General Chemical Corporation exposed 63,000 people to harmful fumes and sent approximately 24,000 people who had inhaled the acid mist to the hospital. Following the release of over 12 tons of acid mist, a poisonous plume developed and drifted over 15 miles from the General Chemical site at Richmond, California. Interestingly, prior to this event, General Chemical conducted a worst-case analysis and estimated a chemical release would produce a toxic cloud that would not travel more than one-half mile from the site. Most of the people who visited the hospital complained of skin rashes and breathing problems, although some people experienced more serious difficulties. As a result of this event, General Chemical was responsible for tens of millions of dollars in legal actions, and public concern over the dangers of plants across the nation became common.

What to Do if Faced with a Chemical Spill

You should be concerned not only with the potential chemical spill within the confines of your business operations, but also with off-site spills at nearby businesses that could affect you as well. A natural gas plant across the street from your facility could experience problems that would directly affect your business, for example.

The obvious starting point is to identify hazardous chemicals kept on site. This requires an audit of janitorial closets, shop areas, mechanical rooms, production floors, and other related areas. While listing the chemicals present, look for properly labeled containers that identify the chemical and the hazards. This is important, especially in secondary containers (small containers and spray bottles). Next, each chemical should have a Material Safety Data Sheet (MSDS) that describes its hazards. By law, this MSDS must be provided to you by the manufacturer or the distributor of the chemical. MSDSs are created for all chemicals, even those that do not present a great danger (i.e., Windex).

> To report a hazardous substance release or oil spill in navigable waters, contact the EPA's national response center at (800) 424-8802. This National Response Center is staffed 24 hours a day amd will monitor the cleanup activities and assist in the cleanup as warranted.

Consider the following helpful hints when developing your chemical spill plan:

- Ask the local fire department to assist you in planning response procedures.
- Ensure that all containers storing chemicals are properly labeled.
- Verify that all chemicals have an MSDS available.
- Keep your MSDS logbook in a central location available to all employees.
- Train employees to properly handle and store chemicals.
- Train employees to recognize and properly handle chemical spills.
- Establish procedures to communicate a spill to employees, management, and the proper local agencies.
- Establish and communicate safe evacuation procedures.

Finally, you should identify other facilities in your area that use, transport, or produce hazardous chemicals. Determine whether chemical spills at their locations could affect your facility, and if it could, address how you would handle that emergency. Think about power plants, refineries, mills, highways, waterways, and railroads. Think worst-case scenario.

If an announcement is made instructing you to "shelter in place," you should not attempt to leave the building. It has been determined that the chemical has been released and exposure is likely to occur if you leave the protective confines of the building. If a shelter in place order is given, here are a few steps you should initiate:

- Close and lock all doors and windows.
- Seal all openings around doors with wet towels.
- Close all fireplace chimney dampers.
- Turn off all HVAC equipment and close or seal all fresh air intakes.
- Seal off, using plastic visqueen and duct tape, all other openings such as vents and exhaust fans.

> For information on EPA's oil spill program, contact the agency's information line at (800) 424-9346. You may also request information via e-mail at oilinfo@epamail.epa.gov. EPA's oil spill reporting website is located at <www.epa.gov/oerrpage/oilspill/contacts.htm>.

Do not attempt to go near and assist victims of a hazardous material spill until the substance has been identified and proper personal protective equipment has been donned or until authorities indicate that it is safe to move about. Doing so prematurely may result in being exposed and endangering your life unnecessarily.

Civil Disturbance and Demonstrations

Facts and Figures

Civil disturbances are part of the nation's landscape. In a land where freedoms abound, citizens have the right to demonstrate and voice their opinions. These demonstrations are usually peaceful. Occasionally they turn ugly and even violent; therefore, it would be prudent to incorporate planning for such possibilities. Civil disturbances encompass not only riots and demonstrations but also such problems as vandalism, looting, gangs, labor disputes, and robbery. In some scenarios, such as labor disputes, the demonstrators are protesting against a certain organization, and the threat to surrounding organizations is small. In other cases, such as rioting after a championship sporting event, nearby buildings and vehicles are exposed to damage and looting. For these reasons it is imperative that you have addressed the possibility of civil disturbance and demonstration.

Case Study 1: Los Angeles Riots, 1992

On April 29, 1992, jurors in Sylmar, California, reached final verdicts in the emotional case involving the 1991 beating of Rodney King by four Los Angels Police Department officers. The entire nation had been able to see the beating, thanks to a citizen who had recorded the entire event on his personal video camera. Many citizens were upset about the brutality of the beating but held back judgment until the case was heard in court. The verdicts were delivered on live television, and the decision surprised and infuriated many. Of the four police officers charged, only one was found guilty of using excessive force. The other three were cleared of all charges.

For many this was another example in a long line of examples of the unjust treatment the LAPD gave the African-American community. Some people demonstrated peacefully, while others began to riot and loot. During the next three days the rioting became particularly violent and innocent bystanders were beaten. One particular example caught on film was the beating of truck driver Reginald Denny.

The mayor imposed a curfew and many schools and businesses were closed. Over 800 buildings were burned and looting was widespread. Parts of Los Angeles resembled a war zone. Governor Pete Wilson called in over 4,000 National Guard troops. Finally, on Monday, May 4, police and the National Guard began to regain control over the city. Schools and businesses reopened, but the scars remained. The entire nation had watched in horror as the events unfolded. The city of Los Angeles received a black eye, the LAPD lost credibility, fifty people were killed, four thousand were injured, twelve thousand were arrested, and $1 billion in property damaged was tallied.

Case Study 2: World Trade Organization Seattle, 1999

The World Trade Organization held its annual conference in Seattle from November 30 through December 4, 1999. What figured to be an event with just passing interest in the news became the story of the week. Tens of thousands of demonstrators showed up to protest commercialism and unregulated growth they believed was encouraged by World Trade Organization activities. The majority of these protestors were peaceful, but as usually happens, a small contingent of troublemakers with nefarious intentions mixed with the crowd. The protestors were actually quite organized. A mixed bag of social activists, environmental advocates, and trade and labor unionists targeted this event, and the numbers of protestors swelled past 50,000. The authorities expected some demonstrations, but the militarism and sheer size of the group left the authorities flat-footed. The Seattle police were unable to establish a safe corridor between the protestors and the Paramount Theatre, where the WTO delegates were gathered.

The crowd became restless and more vocal. The more militant protestors began to set fires and throw rocks and bottles. Somebody tossed a Molotov cocktail. Suddenly a McDonalds restaurant went up in flames. Next, two Starbucks cafes were on fire. The demonstration quickly grew out of control. As the Seattle police attempted to quell the troublemakers, they found the task of separating the peaceful protestors from the bad guys to be difficult. The troublemakers were smart. They made sure to keep peaceful protestors between themselves and the police officers. Earlier, the police had resorted to pepper spray as a means to scatter the crowd; however, now the officers felt stronger force was necessary. As reinforcements arrived, they donned riot gear and used rubber bullets to disperse the crowds. The officers also left the less-effective pepper spray in favor of the more potent CS tear gas.

Mayor Paul Schell declared downtown Seattle a state of civil emergency. He then called on Washington State Governor Gary Locke for help. The Governor called up the National Guard, and a 24-hour curfew was put into effect covering a 46-block area surrounding the WTO conference. Protesting would only be allowed in certain areas away from the conference, and all protestors were required to obtain a permit. As the WTO conference ended and the protestors began to disperse, the city began to calculate the costs. The direct economic impact reached over $12 million for property damage and overtime for police officers. Indirect costs such as lost business and bad publicity certainly cost the city and its businesses millions of dollars more.

What to Do if Faced with a Riot or Demonstration

There are limitations on what can be done during a riot or civil unrest. Taking the law into your own hands can be dangerous and have some very heavy legal ramifications. The best way to mitigate or discourage a riot or vandalism would be to take effective, proactive steps prior to the event. In many scenarios, with a little bit of foresight, trouble can be limited or avoided altogether.

Be aware of events that may trigger demonstrations or unrest. These events might include

- Labor disputes
- Layoffs and downsizing
- Environmentally sensitive meetings or conferences
- Sporting events
- Political rallies
- Economic conferences
- Judicial decisions
- Music concerts
- Religious gatherings
- Biased racial or cultural events

This list is, of course, just a sampling of events that could start out peacefully and turn destructive. Again, do not attempt to solve riot-related events without the assistance of proper law enforcement personnel. If an event is scheduled and you believe it may turn ugly, consult with your local police department. They will be able to advise you in the steps that should be taken. These steps might include

- Hiring temporary guard service
- Installing storm shutters
- Locking gates in the parking lot
- Lowering security grating
- Removing vehicles from the premises
- Removing trash containers or other items that could thrown or set afire

Obviously, it makes sense to distance the demonstrators from the employees of an organization. This is especially important during labor disputes, because those working and those not working will be at odds with one another. It may be advisable to provide the protestors with a place to demonstrate. That may be an off-site area such as an easement or behind a fence or barricade. At Florida State University in Tallahassee, Florida, the university provided local tribal protestors with a fenced-in area in the parking lot. Their protests stemmed from the perceived disrespectful use of the name "Seminoles" and the war chant during sporting events. By providing a place to peacefully demonstrate, the university protected the tribe from harm from counter-demonstrators and protected the university from potential liability.

Another effective way to reduce the threat of civil disturbance involves the use of good public relations. Take, for example, a music and arts festival. Suppose this festival will attract large crowds of college students. Beer will be sold at this venue and the concerts will extend until 2:00 A.M. or 3:00 A.M. Could you foresee a potential problem with an unruly crowd? Certainly, so it makes sense to proactively work with the police department. It might also make sense to contact the concert promoters and seek their assistance. It might even make sense to work with the local colleges and ask them to speak to the students about behavior off campus, reminding them of the police presence in the areas surrounding the festival. These may seem like simplistic ideas that don't really work, but they do.

Loss of Utility

Facts and Figures

Ordinarily, loss of utility might not be considered an emergency. It is an inconvenience, perhaps, but certainly not an emergency or disaster. Consider, however, the plight of a hospital. Loss of electrical utility could be deadly. Consider the call center that serves your local fire department. Without telephone service, they would be unable to respond to emergencies and lives could be lost. In your business, even if lives are not at stake, business continuity is compromised. For this reason you should make certain that your emergency planning addresses the loss of utility.

Case Study 1: New York Blackout, 1977

On July 13, 1977, at approximately 8:37 P.M. lightning knocked out two 345 kV power lines. At approximately 9:41 P.M., lightning strikes caused the loss of two additional lines.

Usually the loss of power can be solved by rerouting power from other nearby utilities; however, because of the loss of protective equipment, all of the nearby utilities quarantined themselves from the Consolidated Edison Electric Company, which served nearly nine million customers in the New York area. The entire electrical system was down from Staten Island, Brooklyn, Queens, and areas north. Some of these customers were without power for 25 hours. It did not take long for some people to take advantage of this opportunity to vandalize and loot area businesses. The looting became so widespread that police were too exhausted and undermanned to pursue the looters. Firefighters were busy as well. Approximately 1,040 fires were attributed to arson during this emergency. After the lights came back on, millions of dollars in damage had occurred.

Case Study 2: California Brownouts, 2001

The most serious energy shortage since the 1970s struck California in 2001. Unusual weather and problems with energy deregulation caused extreme power shortages. Mandatory load shedding plans were activated. Nonessential equipment was required to be shut down. Thermostats were adjusted to 78 degrees Fahrenheit, instead of the normal 72 degrees. These steps helped but did not solve the energy crunch. Finally, a Stage 3 power alert necessitated rolling blackouts across the state. Many employees were forced to work by candlelight. People were stuck in elevators. According to the Electric Power Research Institute (EPRI), a three-day brownout in January cost businesses $1.9 billion in lost productivity.

Confidence in the electric utility industry was lost and the bad feelings expressed by the public increased with the downfall of Enron, one of the nation's top electricity and natural gas marketers. Additionally, people were angry to learn of utility companies' lack of planning and failure to keep up with the demand.

What to Do if Faced with Loss of Utility

The loss of utility is certain in most business operations. The first step in mitigating the damage is to identify utilities that may be affected. These critical operations may include

- Manufacturing equipment
- Computer equipment
- Telecommunications
- Security systems
- Water and sewer
- Gas
- Electric
- Life support systems
- HVAC

Next, determine the impact of service interruption. This impact may be measured in lost production, spoilage, comfort, accessibility, or any other measurable unit. The impact may change as the duration of the interruption increases. For instance, a two-minute power blip may not be cause for concern, whereas a 60-minute interruption may cause great concern. Understand your businesses thresholds for utility interruption.

Finally, establish contacts and procedures for system restoration. This might include contacts with local utility providers, contractors, providers of generator backup, or other related services. In areas of a critical nature, redundancy and off-site/after-hours monitoring should be implemented.

Medical Emergencies

Facts and Figures

Most people spend between eight and ten hours per day at their place of employment. Because of the amount of time spent at work, medical emergencies are a strong possibility. Each year about 1.25 million people experience a heart attack in the United States, and nearly 40 percent of those heart attacks result in death. Heart attacks don't just affect the elderly and retired. Of the 500,000 who die each year from heart attacks, more than 160,000 victims were younger than 65. Imagine an organization that did not plan for a medical emergency such as a heart attack, and then an employee was stricken. Confusion would run rampant. Who would we call? Do we dial 911? How do we get an outside line? What is the address? How will the paramedics gain access to the building? Sadly, during the confusion, the employee might die. However, don't fall into the trap of planning only for heart attacks. That would be a serious oversight. Think about any medical emergency that might occur. Examples might include

- A diabetic experiences seizures
- A lab technician is overcome by toxic fumes
- A line worker's hand gets caught in machinery and the worker loses large amounts of blood
- An employee with high blood pressure experiences a stroke
- A daydreaming employee slips and fall down the stairs
- An employee is walking through the parking lot and trips, breaking a leg
- A nurse is stuck by an contaminated needle

As you can see, the list of possible medical emergencies can be quite long. Some emergencies might be less likely in your particular work environment. A chemical plant that produces toxic substances might be more susceptible to an inhalation emergency. A workforce that includes those with ailments such as diabetes or heart disease should understand the risk and plan for the eventuality of a medical emergency. Consider also the organization that is open to the public and is visited by people with various ailments that are unknown to the host organization. Again, the possibility of a medical emergency of some type is realistic to all organizations and, therefore, should be planned for.

What to Do if Faced with a Medical Emergency

If employees are expected to react properly during a medical emergency, employers must have a procedures manual in place. This manual will describe in detail what events should take place during a medical emergency. For instance

- The names and phone numbers of employees who have emergency medical response training
- Location of the first aid kit or station
- First aid procedures
- Proper disease transmission avoidance
- Injury and illness recordkeeping (OSHA) procedures
- How to reach outside assistance (i.e., dial 911)
- How to get an outside line

The last item is interesting. In many buildings, you must dial an "8" or a "9" to get an outside line. If a medical emergency occurs, confusion and chaos usually follow. We all know we need to dial 911, but in the confusion, we forget to dial the proper suffix for the outside line. Once we dial 911, the dispatcher will ask a number of questions concerning the nature of the emergency, your name, your phone number, and the address where the emergency has occurred. Oftentimes employees or visitors don't know the proper address of the building. Instead, they know a mailing address to a post office box, and the dispatcher will not be able to get emergency response personnel to the building. Finally, give some consideration to how the paramedics will gain access to the building and to the injured. Will they know where to go? Will the receptionist require them to sign in, thus delaying their response? To solve these problems, it is advisable to assign somebody to wait at the building's entrance to escort the paramedics directly to the injured.

> The Red Cross is a leader in medical emergency training. Classes include basic first aid; CPR; ergonomics; slips, trips, and falls; back injury prevention; workplace violence awareness; and managing stress. Typical classes last between one hour and six hours each. You can view course descriptions at <www.redcross.org/services/hss/courses/workplace.html>.

In the event of a medical emergency, employees must have an idea what to do. The procedure manual just discussed will provide employees with such a plan; however, the plan is useless if proper training is not provided. Many options exist, but some of the best training opportunities are delivered by the Red Cross (available at <www.redcross.org>) and the National Safety Council (available at <www.nsc.org/>). These organizations offer local training covering a host of different first aid and medical emergency issues.

Many people would like to help an injured person, but they are concerned that if they do lend assistance they could be sued by the injured party or their family. This creates an obvious dilemma for those normally willing to help. In most states, people who assist victims in medical emergencies are protected by Good Samaritan laws. It is advisable to check with your legal council or local Red Cross office prior to developing your procedures manual. According to the Red Cross, when citizens respond to an emergency and act in a reasonable and prudent way, they will be protected under the Good Samaritan laws. This will protect the rescuer from being sued by the injured person. One caveat does exist though. If a rescuer does offer assistance but is negligent or not properly trained and they further harm the injured person, they could be held liable. Based on this fact, it makes sense to train certain employees in proper rescue techniques including first aid and CPR.

For those working in a healthcare environment, transmission of disease through needles and sharps are a reality that must be addressed. On November 6, 2000, the Needlestick Safety Prevention Act was signed into law. Two months later, on January 18, 2001, OSHA released the final version of the revised Bloodborne Pathogen Standard—29 CFR 1910.1030. Highlights include

- Employers must review the "Exposure Control Plan" annually
- Employers must select safer needle devices as they become available
- Employers must involve employees in the selection process
- Employers must maintain a "Sharps Injury Log." This log identifies which device was being used, the department in which it was used, and how the incident occurred

OSHA estimates that six million workers in the health care industry are at risk of occupational exposure to bloodborne pathogens. Bloodborne pathogens are pathogenic microorganisms that are present in human blood and can cause disease in humans. Possible diseases include Human Immunodeficiency Virus (HIV), Hepatitis B Virus (HBV), Hepatitis C Virus (HCV), and others. Each year, upwards of 800,000 needlestick injuries occur. Many of these injuries are the result of careless work habits such as using a two-handed method to recap needles.[1] In these scenarios, proper training is advised. OSHA also recommends the use of safer needles and sharps, including needleless systems and sharps with built-in safety features that reduce the risk of bloodborne pathogen exposure.

Nuclear Threat and Exposure

Facts and Figures

Fortunately, the odds of being exposed to nuclear energy are not as great as the odds of other types of emergencies. Basically, two distinct nuclear exposures could affect us—nuclear warfare and nuclear exposure from a nuclear power plant accident. Although most people cringe when you mention a nuclear power plant, these plants are actually quite safe. Nuclear power plants are closely monitored and regulated by the Nuclear Regulatory Commission (NRC) which requires each utility that operates a nuclear plant to have both an on-site and off-site emergency response plan. The on-site plans are approved by the NRC, and the off-site plans are evaluated by the Federal Emergency Management Agency (FEMA). These agencies do a very thorough job, and the likelihood of an accident is small. If an accident were to occur and radiation were released, its affects would be determined by the amount of radiation released, the weather condi-

---

[1] When recapping a needle is absolutely necessary, it must be performed using a mechanical device or forceps rather than using a two-handed approach. If a mechanical device or forceps is not available, a one-handed scoop method is performed by holding the sharp or needle in one hand and then scooping up the cap from a flat surface without touching the cap.

tions, the wind direction, and speed. Radiation cannot be detected by the senses. It does not have a smell or a color. It can be detected and measured by sensors.

Even with all the safety measures built into the nuclear power plant, certain people with nefarious intentions can make the seemingly impossible become reality. Given recent world events, we've come to realize that an airplane can be hijacked and flown into a building causing mass destruction. We've heard threats of planes possibly being hijacked and flown into nuclear power plants. And we have the threat of nuclear weapons. Obviously, when considering nuclear weapons, one must understand that guidelines on their use and licensing just don't exist. A rogue nation could, without any warning, detonate a weapon that releases nuclear energy. For these reasons, it is imperative that we include nuclear threat and exposure in our emergency planning efforts, no matter how remote we may believe the threat to be.

Case Study 1: Three Mile Island, 1979

Historically, the United States nuclear power industry has been a model of safety and efficiency. That perception was seriously and perhaps permanently damaged when the Three Mile Island Unit 2 Nuclear Generating Station in Harrisburg, Pennsylvania, was damaged. This event led to the most serious and infamous commercial accident in U.S. history. The nuclear industry would never be the same. Changes to standard operating procedures, safety contingencies, and regulations have been introduced, yet we all know the seriousness of the threat of a nuclear accident.

On March 28, 1979, at 4:00 A.M., the system which feeds water to cool the steam generators malfunctioned. Emergency systems began operation, but a valve failed to completely close resulting in a small loss of coolant. This event was further compounded by operators who misinterpreted the developing problem. The minor mishap progressed into a full-blown accident that caused substantial damage to the reactor and associated components. All of the fuel cells and over 90 percent of the reactor core were damaged. The containment building in which the reactor was located as well as several other locations around the plant were contaminated. Despite the severity of the damage, no injuries from radiation occurred. The resulting contamination led to the development of a ten-year cleanup and scientific study, and today the reactor is inactive.

What to Do if Faced with Nuclear Exposure

The Nuclear Regulatory Commission requires that nuclear power plants have a system for notifying the public in the event of an emergency. There are four standard classifications of emergencies:

- Notification of Unusual Event—This is a non-specific warning and is the least serious of the four categories. The event poses no danger to employees or the public and no action (i.e., evacuation) is required of the public.

- Alert—This classification is declared when an event has occurred that could jeopardize the plant's safety, but backup systems are in place. Emergency agencies are notified, but no action is required from the public.
- Site Area Emergency—This classification is declared when an event has caused a major problem with the plant's safety system and has progressed to a point that a release of radiation into the air or water is possible but would not exceed regulations instituted by the Environmental Protection Agency. No action is required by the public.
- General Emergency—This classification is the most serious and comes when the plant's safety systems have been lost. Radiation could be released beyond the boundaries of the plant. Emergency sirens will sound, and some people will be evacuated or instructed to remain sheltered in place. Listen for more specific instructions.

If you receive an alert, remember

- A siren or tone alert does not necessarily mean you should evacuate. Listen to the television or radio for further instructions.
- Do not call 911. If this is a true alert, a special rumor-control phone number will be provided.
- If you are instructed to shelter in place, be sure to close all doors, windows, and chimney dampers, and turn off all HVAC equipment.

If you receive a warning and you are instructed to go inside, it is advisable to shower and change clothes. After removing your clothes and shoes, place them in a plastic bag and seal the bag.

Typically there are three ways to minimize radiation exposure:

- Time—Limit the time you spend near radiation. This seems obvious, but it's worth repeating. During an exposure event, local authorities will monitor the radiation, which should decrease over time. They will inform the public when the danger has passed.
- Distance—The farther you are away from the source of radiation the better. For this reason, local authorities will likely evacuate those near the radiation release.
- Shielding—Human skin is very sensitive and does little to insulate the internal organs from the effects of radiation. Other materials such as concrete and steel in buildings, for instance, may provide some shield from radiation for short durations. For this reason, you may be asked to shelter in place.

Structural Collapse

Facts and Figures

Fortunately, the collapse of a building is a very rare occurrence. Credit agencies such as OSHA and the existence of building codes for ensuring safe construction practices and structural integrity. Unfortunately, when a structural collapse does occur, it is always costly, both in property damage and human life. Structural collapse is usually the result of an unusual external event such as bomb, fire, or—as we've witnessed—a fully fueled jumbo jet being flown into the building. The next two case studies concern faulty construction.

In the early 1980s, various municipalities created specialized urban search and rescue teams to assist with emergencies, especially those involving structural collapse. These teams gained popularity as their successes were documented. Their expertise has been called upon both domestically and internationally. In 1991, the Federal Emergency Management Agency (FEMA) incorporated search and rescue teams into the Federal Response Plan. Currently FEMA sponsors over 25 national urban search-and-rescue task forces. These task forces are equipped to conduct search-and-rescue operations following earthquakes, tornadoes, floods, hurricanes, aircraft accidents, hazardous materials spills, and catastrophic structure collapses.

Case Study 1: Point Pleasant Cooling Tower Collapse, 1978

April 27, 1978, began like many others at the Point Pleasant power plant. By 7:00 A.M. tradespeople were busy working on one of the two cooling towers being erected on site. Suddenly and without warning, one of the two towers began to give way. In a matter of seconds the tower collapsed and 51 lives ended. Workers on the ground compared the helpless eventuality of the falling tower to dominos. Safety nets had been placed in the mouth of the tower, but they provided no purpose at all. OSHA arrived on the site and conducted investigations. Final results pointed to "green concrete" as a major cause of the collapse. As the tower was being erected, sections were formed up and concrete poured. While the concrete was still green, or not fully cured, the forms were stripped off and the next section was formed and readied for the next pour. Additional blame was directed at scaffolding that surrounded the structure. As the tower became taller, so did the scaffolding. The massive scaffolding system was held in place by bolting into the exterior wall of the cooling tower. When these reinforcing rods were placed in areas of concrete that were not fully cured, they put undue stress on the structure. In hindsight, this was an obvious recipe for disaster. Since this failure, OSHA has created new concrete testing standards. OSHA calculates these new standards will cost contractors approximately $28 million per year. Although the expense is high, OSHA believes it will save an average of 27 lives per year; a reward that we can all agree is worth the price.

Case Study 2: Hyatt Regency Walkway Collapse, 1981

On July 17, 1981, guests staying at the Hyatt Regency in Kansas City were victims of a very sad and deadly catastrophe. The newly built hotel was hosting a dance contest that attracted nearly 2,000 people. Many of these guests were on the ground level of the hotel lobby while others were dancing on the hotel's suspended walkways. These walkways connected the hotel's conference room facilities on the second, third, and fourth floors. A design flaw caused the suspended walkways to collapse, and 114 people died while hundreds more were seriously injured. The collapse occurred because the walkways were not built to withstand the load. During the design stage a miscommunication occurred between the structural engineer and the company hired to fabricate and erect the structure. Upon completion of the investigation the engineering company's license was revoked and the engineers who had affixed their seals to the design drawings lost their licenses in the states of Missouri and Texas.

What to Do if Faced with a Structural Collapse

Because of the specialized nature of search and recovery and the danger inherent to that type of work, it is advisable to bring in professionals to coordinate rescue efforts when a building collapse has occurred. FEMA has developed and can deploy teams of trained response personnel to coordinate your activities.

If an emergency occurs and warrants search and rescue support, FEMA will deploy the three closest search and rescue teams within six hours of notification. If additional teams are necessary, FEMA will provide them as well. These teams serve to support state and local emergency responders.

> Two organizations are leaders in structural collapse education and response. The National Association for Search and Rescue (available at <www.nasar.org>) and Federal Emergency Management Agency (available at <www.fema.org>) are excellent resources for the unique challenges facing a structural collapse.

There are 62 people, divided into two 31-person task teams, in each FEMA task force. Each 31-person task team includes four canines and 60,000 pounds of equipment valued at over $1.4 million. The task teams are totally self-sufficient for the first 72 hours of their deployment.

Task force members are highly trained and specialized professionals. They do much more than simply provide the labor for digging through the rubble of a collapsed building. Each task force member is a trained emergency medical technician and is required to complete hundreds of hours of specialized training. Typically, the task force team is divided into four unique areas of specialization:

- Search team—Finds victims trapped after a disaster
- Rescue team—Safely extricates victims from the rubble

- Technical team—Structural specialists who evaluate and stabilize the structure
- Medical team—Cares for the victims of the disaster

The FEMA urban search and rescue teams are not limited to building collapse. They can be deployed after mining accidents; transportation accidents; and natural emergencies such as hurricanes, tornadoes, earthquakes, floods, hazardous chemical releases, and terrorist activities. Additionally, FEMA provides training to first responders in search and rescue techniques and may be able to provide grants local communities.

Workplace Violence

Facts and Figures

Unfortunately today we must be aware of the possibility of workplace violence. Certain industries are more susceptible to violence than others. Certain locations create a more dangerous and volatile atmosphere than other locations. But don't fall into the trap of believing your business is immune to workplace violence.

What is workplace violence? It can be offensive or threatening language. It may be a threatening note. It can be a push or a shove. It may be a mugging, rape, or robbery. In its most extreme form, it can be homicide. In fact, homicide is the third leading cause of fatal occupational injury in the United States.

According to a Bureau of Labor Statistics census, in 2000 there were 674 workplace homicides. These homicides accounted for 11 percent of all the fatal work injuries in the United States. Retail industry accounts for the greatest number of workplace homicides; in fact, retail typically accounts for half of the recorded workplace homicides each year. Not surprisingly, robbery is the leading motive for workplace violence. Those working in healthcare, government, and judicial fields are also especially susceptible to violence.

Factors which may increase a worker's risk for workplace assault might include the following:
- Handling money
- Serving alcohol
- Working in high-crime areas
- Driving delivery vehicles
- Having a mobile workplace such as taxicab drivers or meter readers
- Working in law enforcement
- Working with unstable or volatile persons in health care, social services, or criminal justice settings
- Working alone
- Working late at night or during early morning hours
- Guarding valuable property

Case Study 1: Royal Oak Postal Facility, 1991

On November 8, 1991, letter carrier Thomas McIlvane, lost his appeal for reinstatement to his job at the Royal Oak post office just outside of Detroit. He had been fired for alleged insubordination stemming from false statements recorded on his timesheet. He vowed to "get back" at his supervisors. Just six days later he stormed through the loading dock area, passed through two large double doors, immediately grabbed a woman, and held his sawed-off .22 caliber Ruger rifle to her head. He let her go saying, "You're not the one I want." As McIlvane entered the office area of the postal facility, he searched for specific people. In ten minutes, he shot ten postal employees and killed three including the former supervisor who turned down his appeal for reinstatement. Two other employees were injured as they jumped out of windows attempting to flee the shooting. After satisfying his anger, he turned his rifle on himself and pulled the trigger, instantly killing himself. McIlvane, a 31-year old former Marine, was described by former co-workers as being a "waiting time bomb" and violent.

Case Study 2: Columbine High School, 1999

On April 20, 1999, the country watched in horror as the news report of yet another school shooting came across the wire. This particular event, though, proved to be especially tragic. At about 11:19 A.M., two Columbine High School seniors, Eric Harris and Dylan Klebold, began a killing rampage that killed fifteen people and injured dozens more. The gunmen began their attack in the school parking lot and slowly made their way into the school's hallways, cafeteria, and finally the school library. By 11:35 A.M. the gunmen had wounded their last victims. By shortly after noon, they had committed suicide.

The attack was gruesome enough, but as the investigation ensued in the coming weeks and months, the public learned of the true intentions of these killers. Harris and Klebold had intended to kill far more people than they actually did. They had successfully planted a pair of 20-pound propane bombs in the school cafeteria. Had these bombs detonated they would have certainly killed the nearly 500 people in the cafeteria. Additionally, Harris and Klebold had planned to leave the school, hijack a plane, and fly it into a building. The public was shocked to learn of the depth of the anger of these two young men.

What to Do if Faced with Workplace Violence

OSHA requires that employers address workplace violence issues. Specific regulations have not been created, but OSHA's General Duty Clause covers all safety issues, such as ergonomics, indoor air quality, and workplace violence. OSHA requires that employers initiate a safe and healthy workplace and

- Identify hazards in the workplace

- Create a written safety plan
- Perform regular safety inspections
- Provide safety and health training
- Record and report injury and illness
- Commit to safety
- Ensure employee participation
- Evaluate the program periodically

OSHA believes that employers who enact the preceding requirements will proactively address the potential for workplace violence. First, employers must understand the seriousness of this problem. Management must take this problem seriously and must be committed to a safe workplace, not just in word but also in deed. This may require new human resource policies such as "zero tolerance." Secondly, employers must examine both historical accounts of workplace violence and the future possibility of such actions. If trends are identified, enact and enforce rules to mitigate these problems. Finally, employers must train their employees how to identify behavior that may indicate a tendency towards violence and how to handle a volatile situation. These three core responsibilities will provide a solid foundation for workplace violence mitigation.

Often behavior recognition and environmental controls are important factors to consider when attempting to reduce occurrences of workplace violence.

Behavior solutions might include
- Don't give orders.
- Remain calm.
- Show a caring attitude.
- Don't touch the other person.
- Don't raise your voice.
- Don't match the threats.
- Use calming language such as "I know you're frustrated. Let me try to help."

Environmental solutions might include
- Provide security (or better security).
- Install metal detectors, cameras, or security personnel.
- Restrict areas that the public can enter.
- Do not allow the public to enter the office area of the building.
- Provide alarms (verbal, mechanical, and/or visual) to warn of danger.
- Provide adequate illumination at entryways, parking lots, corridors, and other areas.

> Section 5 (a)(1) of the William Steiger Occupational Safety and Health Act of 1970 is known as the "General Duty Clause." This section reads as follows:
>
> 5. Duties
>
> (a) Each Employer
>
> (1) Shall furnish to each of his employees employment and a place of employment which are free from recognized hazards that are causing or are likely to cause death or serious physical harm to his employees.

- Provide escorts to the parking lot at night.
- Ensure that employees are aware of the location of emergency exits.
- Install deep transaction counters to limit physical contact with the public.

In the event of a workplace emergency, first responders such as the police and emergency medical technicians need to know exactly where to go to assist the victims and apprehend the suspects. These responders are unlikely to be familiar with your building; therefore, you should assist them by having this information available prior to their entry into the building. In the Columbine High School emergency, floor plans were not available to the first responders. Students had to draw crude floor plans from memory so the responders could decide on the most appropriate action to take.

The Greatest Disaster That Never Happened

Like many people, I eagerly awaited the new millennium. I remember sitting in school as a young boy and calculating how old I would be in the year 2000. At the time, it seemed so distant, too far away to give serious consideration. Now I was faced with the reality of this event. As an adult with adult concerns, I was tasked with the responsibility to make sure "Y2K" didn't shut the business down. It seems that when many computer programs were created, the programmers were too short-sighted to realize that computers wouldn't be able to distinguish "00" from 1900 or 2000, and with this slight misunderstanding, events of apocalyptic proportions were predicted to occur.

If we were to listen to the doomsayers, when the clock struck midnight, all bank account databases would lose their data, elevators would quit working, electricity would no longer flow, air traffic control would quit operating, and so on. We were told to gas up our cars, withdraw cash from the bank, hoard food, and purchase guns!

Thankfully, those events failed to occur, in large part due to the concerted effort to fix the problem before it happened. Would Y2K have been as cataclysmic as predicted if not for the contingency planning and de-bugging done? I don't know, but I do applaud the technology gurus who kept us from potential disaster.

I do believe that some positive came from this Y2K exercise. It forced us to examine our workflow and develop contingency plans in the event everything did stop working. These exercises are the classic disaster planning models that we use for hurricanes, floods, tornadoes, riots, and other such disasters.

Summary

Given the harsh realities of today's world, non-natural emergencies such as terrorism, bomb threats, civil unrest, and nuclear and biological concerns are of great concern. Although many of these threats may appear to be outside of the scope of our expertise, steps can be taken to mitigate the negative impact. Tools such as the ATF's bomb threat checklist can provide the authorities with information that could help locate the perpetrator. Learning proper bomb sweeping techniques can be critical if local police departments are unable to quickly offer assistance to your business.

Even if your business never experiences a bomb threat or a terrorist act, you will likely be faced with unforeseen medical emergencies. Employees or visitors could be affected, and how quickly you are able to respond to their needs could be the difference between life and death. For these reasons, it is imperative that you plan for the eventuality of non-natural emergencies.

4

The Basic Stages of Planning

Where Do I Start?

When creating an emergency plan, your success starts with your ability to understand the finished product. What will the plan look like? What will it cover? Whom will it address? It is critical that the driver of this project have a clear understanding of these issues prior to commencement. At some point, however, you must change your focus from the end product to the start of the project. Where do I start? How do I sell this to management? Where will I get the resources? All very good questions, which we will discuss shortly.

Breaking the Large Task into Bite-Sized Pieces

Any time projects of any size are undertaken, project managers must not let the magnitude of the projects freeze them. Do your research; think about the finished product, but at some point the pen must meet the paper. Failure to do so results in paralysis by analysis, and you're not likely to get acceptance from management if you find yourself caught in this web of indecision.

Ask a project manager for advice and they will always encourage you to break the large, seemingly overwhelming project into bite-sized and workable pieces. Excellent advice whether building a skyscraper, installing a large computer network system, or creating an emergency management plan for your company. The logical starting point is to first determine your objectives and scope.

Determining Your Objectives and Scope

You cannot begin a journey until you have an idea where you'd like to end up. Well, I take that back, sometimes you can, but in this journey, you're on the company payroll and the boss is unlikely to finance a project without direction or an apparent end. To identify where you hope to be, ask a few simple questions:

- What do I want this plan to do?
- Why do we need a plan?

- Who will this plan have jurisdiction over?
- What events present the greatest liability?

These questions form the basis for your emergency planning project. They identify both objective (specific event versus all emergencies) and scope (departmental plan versus company-wide plan). Until you ask these key questions, you really can't to the next step. What is the next step? Now would be a great time to identify your resources, either members of your team or those business experts who can help you put the plan together.

Identifying Your Resources

Yes, a team of people will be needed to adequately develop a plan. Even if you are a small organization, you'll need the expertise of a diverse group of people. Regardless of your title and position, you can't possibly know all of the aspects of your company in the detail required for a workable plan. Your knowledge base may be one of great breadth (and that's a great attribute for the emergency planner to have), but you'll need the assistance of people whose knowledge has real depth to it. Having a multi-disciplined approach to the emergency plan is of utmost importance.

Understanding the Costs Involved

Depending on the scope of your emergency plan, costs can quickly escalate. You have an obligation to management to estimate the cost prior to commencement. It's not likely, nor is it advised, that you simply purchase a book (this one included) and copy the plan without adjusting it to your organization's specific needs and idiosyncrasies. If the development of a plan were that easy, the costs would be low, the process not labor intensive, and everybody would do it tomorrow. For most organizations, the cost need not be large. Here are some of the costs with which you are likely to be faced:

- Non-billable time while producing the plan
- Support from administration (typing, editing, and copying)
- Printing the emergency plan
- Distribution of the plan
- Training and drills
- Revisions to the plan

Again, this book is not a one-size fits all solution to emergency planning. Beware of software and other resources that over-promise and under-perform. What about this book? This book is a great vehicle for educating the average employee about the steps involved in developing a workable emergency action plan. Boilerplate audits, checklists, and even a sample emergency plan are included in electronic format so you can edit and customize the documents for your own use.

Presenting Your Case to Management

One area of great concern is the role of management. Many a good planning team has been disbanded and told to "Get back to work" because management failed to recognize the importance of the emergency planning task. Certainly in light of September 11, emergency planning will gain greater acceptance, but what if an event of that magnitude never occurs again? Will we forget? Will we relegate emergency planning to a secondary or even tertiary concern? Knowing human nature, it could happen, and it will be up to you to ensure that this lapse does not occur in your organization.

Justifying the Cost

If management balks at the cost or the time required to effectively plan for emergencies, you must take a stand. In any task of this magnitude it is absolutely key that management be aware of and supportive of your vision. If fact, the vision must be a shared one. This may take a bit of convincing or at least a good dose of resource reality. You don't wake up one day and decide you want an emergency plan, go to the store, and buy one. It just doesn't work that way. Each organization is unique. The geographic location of your organization introduces unique challenges. The business you conduct presents its own special challenges.

Give management an idea of the resources (time, funding, and personnel) that you'll need to complete this task. Too often, those employees assigned the task of providing an emergency plan treat it as something other than a real project. Regardless of your profession, you have been assigned a very important and complex task. Treat it as such. Develop a scope. Create a schedule. Do a cost analysis. Write a mission statement. This is a real project with real objectives and milestones. By treating this project as any other important project that you undertake, you will instill an attitude of seriousness and professionalism in your planning team and in the management team that you are attempting to persuade.

Consequences Versus Rewards

Typically there exist two avenues of thought on selling your ideas to another person. One tactic is to stress the negative impact of not following a recommended course.

Negative selling points
- Inability to quickly rebound after an emergency
- Financial loss
- Loss of market share
- Higher insurance premiums
- Bad press coverage
- Increased chance of legal liability

The consequences of not having an emergency plan are costly. Fear serves as a great motivator, so tactfully using fear and consequences is an effective way to highlight the importance of having an emergency and business continuity plan.

A second tactic is to stress the positives of following a recommended course. By appealing to management's sense of responsibility or greed (or both), you can present quite an effective case for your desired course of action. Everybody likes to be rewarded. If you can demonstrate effective ways to reduce risk and therefore protect the company assets, you increase your chances of persuading management to accept your proposal for emergency planning.

Positive selling points
- Moral responsibility to employees
- Better able to rebound financially
- Compliance with codes and regulations
- Business continuity
- Possible reduction in insurance premiums
- Reduces civil or criminal liability
- Environmentally responsible

> In his excellent book *Artful Persuasion: How to Command Attention, Change Minds, and Influence People*, Harry Mills explains "people are more motivated by the fear of losing something than by the reward of gaining something of equal value. Psychologically, it's much more painful to lose $100 than it is pleasurable to win $100."[1]

When deciding upon the use of rewards or consequences, I do not believe that you should focus on one area. A person skilled in persuasion will use both tactics to influence his or her audience. If your corporate culture is such that management is particularly persuaded by rewards, then spend more time outlining the rewards. In doing so you'll capture their attention, but don't forget to discuss the negative consequences of not having a plan as well.

The Mission Statement

Most likely you are familiar with the mission statement. It seems that the organizational philosophy of the 1980s dictated that all companies have one.

Organizations both large and small have created mission statements to articulate who they are, what they do, and what they intend to do in the future. The mission statement has become the cornerstone of many good organizations. By the same token, for many, the mission statement has become an overused yet underutilized formality that holds little bearing on a company and its business.

Although some people may discount the importance of the mission statement, I believe the mission statement does serve a worthwhile purpose, especially in the early stages of

---

[1] Harry Mills, *Artful Persuasion: How to Command Attention, Change Minds, and Influence People*. Amacom, 2000, p. 127.

any large new venture, including emergency planning. In these formative stages, many people, especially management, will not fully understand the scope or even the purpose of emergency planning. Sure, it sounds important. We know it's the right thing to do, but what will it do for us? Who will be covered? The mission of the mission statement is to answer those questions in thirty seconds or less. Think of the mission statement as an executive summary of sorts.

Typically, the mission statement answers the following three questions:
- What we will do?
- Why are we doing it?
- How we will do it?

> **Emergency Management Team Mission Statement**
>
> Our mission is to protect the employees, property, business, and the community in the event of an emergency.
>
> We are committed to providing the company with effective and useful solutions that will reduce or eliminate emergencies and their negative impact.
>
> We will accomplish this by conducting risk assessments, creating a written emergency action plan, and conducting employee training.
>
> Additionally, we will create viable plans for restoration and recovery in the event of an emergency.

The mission statement actually serves three distinct purposes, each independent and each occurring at different stages on the timeline.

In its beginning stages, the mission statement serves as a home base for the emergency management team. If the team is ever unclear about its purpose or scope, they can go back to the mission statement and get back on track. Secondly, it serves as an effective marketing tool for the proposal to management. It tells them exactly what we plan to do and why we are doing it. It serves to justify the endeavor, which will require the support and resources of management. Lastly, the mission statement bestows legitimacy to our endeavor. This plan must be accepted and supported by all employees, from the CEO to the night janitor.

For obvious reasons, the mission statement is most effective when created by the emergency management team. However, at some point, ownership of the mission statement should be transferred to the organization's corporate executives. This must happen prior to the mission statement being distributed to the rest of the organization. A mission statement coming from the ranks of management carries much more weight than a mission statement created by a team consisting of various mid-level managers and front-line workers.

The emergency management team must have authority to make decisions and procure resources. Corporate management can facilitate those needs by heralding the mission statement as its own.

Establishing Authority and Chain of Command

An obvious area of concern and the cause of many planning failures stems from a lack of authority. The position of emergency planning director is a position that should not be granted without careful thought. The elected individual should be sensible, able to direct and delegate, able to communicate with all levels of the organization (not an easy task), and—most importantly—able to make decisions without fear of being vetoed by others. As we mentioned, effective emergency planning requires the input of a very diverse group of people. The emergency planning director should have the ability and the courage understand different points of view and make a decision that is best for the organization. As we well know, any time a group is formed, different views and ideas will materialize. The planning director needs to have authority to make a competent decision based on the data received.

Understanding Your Business

The Occupational Safety and Health Act requires employers to evaluate the hazards that exist at their workplace and create an emergency plan specific to their business. For instance, a healthcare facility worker will be exposed to hazards that office workers would not be. Exposure to bloodborne pathogens, needlesticks, and sharps are a very real concern for healthcare workers and must be addressed in the corporate safety plan. Browse though a corporate safety plan for a retail store and its unlikely that needle sticks will be addressed. Its not a concern; those hazards do not exist in that business. OSHA does not provide businesses with a "canned" safety plan because every business is unique and the exposure to hazards is different. The OSH Act requires that you evaluate your business and develop a customized plan to keep your employees safe. Following this same line of reasoning, you need to evaluate your business and the hazards that are produced by your building use and develop an emergency plan that addresses those hazards. Where is the property located? In a flood zone perhaps? Do you own the building or is it leased? What kind of business are you conducting? What functions of your organization are most critical to its survival? These are the types of questions that must be asked when developing your emergency action plan. To develop a useful emergency plan, you must understand your business.

Identifying Core Business

In any business there exists core functions that define the organization. For instance, an electrical utility company has a core function to provide electricity to customers. You can break the customers into priorities as well. A utility company may decide to restore power to hospitals first, law enforcement agencies second, and eventually residential neighborhoods. Each organization is encouraged to identify their core business and then examine ways to protect that core business in the event of an emergency. In the event business is interrupted, you must have some idea of what aspects to restore first. If this important step is not completed, you risk spending valuable time restoring non-core business functions. All resources available must be used to restore business continuity.

Hospitals are an example of an organization expected to provide core functions during and after emergencies. To ensure that this happens, the Joint Commission on Accreditation of Healthcare Organizations (JCAHO) performs regular inspections of healthcare organizations. During these inspections the Joint Commission reviews items such as the generator capacity, maintenance records, water storage, and bioterrorism plans. These inspections can be grueling and quite intensive, but they are also very necessary. We understand that hospitals serve a valuable and necessary function. We understand that without hospitals, recovery from an emergency would be difficult, maybe even impossible. If your business is not part of the healthcare industry, you might not have an external agency watching your every move. If that is the case, don't feel like you can relax. You should create an audit process to ensure that your organization is prepared in the event of an emergency. This discussion should start with the identification of your core business and quickly move to the identification of the cost of business interruption.

Cost of Business Interruption

After many emergencies organizations have difficulty recovering. Any delay in production of a product or delivery of a service will have a negative effect on the cash flow of the business. Some companies have reserves or are able to conduct business in a remote area not affected by the emergency; however, many business large and small never recover and are forced to close their doors. After the 1993 World Trade Center bombing, nearly half of the businesses located in the towers went out of business as a result of the bombing. That number is alarming! Furthermore, a very small percentage of those organizations had a business continuity plan in place. For this reason, the cost of business continuity must be examined. Here are a few of the questions that need to be answered at a strategic level:

- What is the daily cost of unproductivity?
- How long can we absorb a business interruption?
- Will the inability to provide a product or service affect our future market share?
- Will the inability to provide a product or service affect our reputation in the industry?
- Does it make sense to acquire insurance to protect lost productivity?
- Does it make sense to create financial reserves for business continuity?

- Is the cost of flexible, contingent office space a viable option?
- If we are leasing the space, what assurances do we have that the property owner will move quickly to repair the property?

These questions must be answered at an executive level within the organization. Most managers, including those responsible for emergency planning, understand how their department supports the organization but are less likely to understand how other business units and departments support the organization. Emergency planning managers need to understand the organization from a point of view that extends beyond the specific area they control. An organization will often identify critical mission or business functions in a business plan. Even with a business plan in place, it is unlikely that all of the answers will be readily available; therefore, research must be done. Options must be examined against three factors—probability of the emergency, liability of the loss of productivity, and cost. There must be a worthwhile return on investment.

Understanding Your Property

When creating an emergency action plan for your business, a number of issues play a critical role in the events you are likely to experience. While you can't account for every emergency event, you can address those that present the greatest risk. Later in this chapter we will explore the risk assessment matrix as a tool that can help us to identify the events that pose the greatest liability. Based on the results of the risk assessment, the emergency planner would begin to develop step-by-step procedures for the emergency events. Before we delve into the actual risk assessment we should spend some time examining the various factors that could increase our exposure to an emergency event.

Leased Versus Owned

The relationship between lessor and lessee can create a number of emergency management questions. The tenant may feel that total responsibility for emergency planning and recovery belongs to the landlord. In turn, the landlord may expect each tenant to be responsible for its own emergency planning. This disconnect is often not realized until its too late. To prevent misunderstandings, at the time of lease signing and at least annually thereafter, the owner of the property should meet with representatives of the tenants to review the emergency plan. This would be a great time to schedule evacuation drills and elect floor captains. It is critical that a chain of command be established. One person should be elected as the emergency manager for the entire building. Usually this person is a member of the landlord's team because they will be making decisions that affect the whole building. This person should be assisted by a team of representatives from each tenant. Among the core functions of the emergency manager is the communication of proper evacuation procedures to the team members. These team members would be responsible for training their individual employees. Additionally, the team members would keep the emergency manager informed of changes that might affect the safety of the building. Examples of such changes include chemicals stored, employee count, and floor plan reconfigurations. Additionally, here are additional questions that should be considered:

- Who is responsible for staging evacuation drills?
- Is each tenant responsible for its own emergency plan?
- Who makes the decision to return to the building after an evacuation?
- What happens to the lease if the property is uninhabitable for an extended period of time?
- What kinds of insurance policies are required and in what amounts?
- Are the landlord, tenant, or both parties required to acquire insurance?
- How much damage must be assessed before the property is considered uninhabitable by the insurance provider?

Geographic Location

The geographic location of your building is perhaps the greatest indicator of the types of emergencies for which you must prepare. Those businesses stationed on the eastern seaboard, particularly from the Carolinas south, understand the probability of a hurricane or a violent tropical storm. Based on their geographic location, organizations in Hawaii, Alaska, Washington, Oregon, and Northern California should understand the dangers of volcanoes. Companies in the Midwest realize the likelihood of a tornado. It seems that each area in the country is susceptible to its own unique emergency events; if we live in a certain area, we understand these hazards and prepare for them accordingly. This is where the real danger lies. If we are not careful, we will focus on the obvious events that could occur and overlook the unusual—but very possible—emergency events. For instance, it has snowed in Florida. Terrorist bombs have rocked the quiet Midwest. Floods have occurred in the desert, and earthquakes can strike in Boston, Massachusetts.

When performing a risk analysis for a building based on its geographic location, we sometimes focus on the natural, weather-related events. True, these events are more common and the damage is often more costly, but do not forget to analyze all emergency events that could occur. For instance, is your building located near a nuclear power plant, refinery, or a chemical manufacturing plant? If it is, you should have a specific plan in place to deal with these potential liabilities. This isn't news to most of us. We realize the dangers that exist when located in close proximity to these types of businesses; but let's take this example a bit further. Is your building located near a highway, railway, or navigable waterway? Could this transportation route be used to transport hazardous chemicals? Suppose an accident occurred and hazardous chemicals were released. Suddenly, a cloud of poisonous gas drifts toward your property, located two miles away. Would your organization know how the handle the emergency? This example shows the importance of going beneath the surface, really digging down and finding the not-so-obvious dangers that are present.

When examining our business, we may feel insulated from many emergency events because we are a small, low-key organization. We don't produce hazardous chemicals, we believe in a safe workplace, and historically our city has not experienced many devastating emergency events. This may be true, but based on proximity to other businesses and events, dangerous situations could materialize. The risk factors outside of your immediate control might include

- Sporting events
- Political rallies
- Flood zones
- Water dams
- Large metropolitan areas
- Airports

In a multi-tenant environment, the business being conducted by other tenants can have a negative impact on your organization and employees. In fact, in some scenarios your business continuity or even the lives of your employees are in danger. For example, abortion clinics are frequently the target of demonstrators. If your company shares space with an organization with a bad public image, you could have problems. Issues range from impeded access because of demonstrators, to verbal abuse, vandalism, and even injury or death. If this is a concern for your organization, you must address the problem. It will be necessary to distance your company from the target organization. Speak with the organizers of the demonstrations and express your concern as an uninvolved third party. If they are legally demonstrating, you have no recourse but to try to make the best of a bad situation. Perhaps identification can be worn that would distance you from the target. Stress your organization's desire to remain neutral in such affairs.

The situation is much different if your organization is the target of such demonstrations. You must take a different approach. Reasoning with the demonstrators will probably not work. Fighting back will definitely not work. In many similar situations, the target company has hired security guards to protect those crossing a picket line or entering a clinic. These organizations have found that an unbiased third party with a badge and a gun can keep relative peace. The installation of security cameras may influence the crowd not to do anything radical. The installation of fences and gates are often used to keep a buffer between the demonstrators and the individual targets. Sometimes employees who cross a picket line are bused in with a police escort. Visitors to a company that is the target of protests may be required to use a back entrance that has restricted access. Regardless of the situation and the eventual solution, you must address these possibilities in your emergency action plan.

So where is your property located? Does that location introduce factors that could create an emergency? What are the obvious issues that endanger your business? Don't focus solely on the obvious, but explore all possibilities. Don't limit your exposure to weather-related or naturally occurring events. Understand that many emergencies can be the result of human error.

Building Use

Consider the kind of property you are managing and the tasks that are performed there. Each property will have its own unique issues and concerns. An aware emergency planner would have a good understanding of these issues and would customize the emergency plan to address these issues. The planner would also consider the building occupants and their attitudes or physical and mental limitations. Based on the planner's observations, details would be added to the emergency plan that address these issues as well. A number of different building classifications exist, but we are going to examine just four types—the industrial, commercial, educational, and healthcare facilities.

- Industrial Facility: Industrial facilities tend to house equipment and chemicals that could produce hazards if used improperly. Chemical spills and fires are not uncommon in such buildings and, therefore, must be addressed in the emergency plan. Questions to ask include
    - Is an industrial fire brigade necessary?
    - Can we contain and cleanup a chemical spill?
    - Is the alarm audible over the machinery?
    - Have all shifts been trained in emergency procedures?
    - Have employees been trained in shut-down procedures in the event of an evacuation?
    - Do we have a way to safely store flammable and toxic chemicals?
- Commercial Property: Multi-story, multi-tenant commercial buildings present a number of communication and logistical problems when emergency planning is concerned. Too often commercial tenants view emergency procedures and evacuation drills as unnecessary and a waste of time. The emergency planner should be aware of these attitudes and work toward changing these perceptions. Questions to ask include
    - How would all occupants be notified?
    - Has a chain of command been specified?
    - Have all tenants been trained in emergency evacuation procedures?
    - What kinds of flammable or toxic chemicals are the tenants storing?

- Do the chemicals all have current MSDS documentation?
- Have emergency responsibilities and expectations been added to the lease agreement?
- Who is responsible for testing alarms, fire extinguishers, and other related equipment?
- Does the emergency planner have after-hours phone numbers for the tenants?
- If tenants make changes to the building structure or floor space, have the changes been added to the as-built drawings?

- Educational Facility: Educational facilities present a number of challenging issues, and each of those should be addressed in the emergency plan. Pre-school and elementary schools are not likely to have many of the hazards that higher grade educational facilities might have, but the lower comprehension level of the students must be addressed. High schools, technical schools, and colleges will have labs that use chemicals and machinery that could malfunction and create an emergency event. Evaluation of facilities should take into account both the hazards that exist and the comprehension level of the students. Because of the young age of the majority of the occupants in an educational facility, the administration must provide an increased level of direction and hands-on assistance when evacuations are required. Questions to ask include
  - Have we established specific staging areas for each classroom?
  - How will we handle bomb threats?
  - How will we handle false alarms?
  - How will we communicate with parents?
  - How will we account for students who are picked up by parents?
  - Can transportation be brought in to take students to a safe location?
  - Have we addressed the potential for violence?
  - Is the school facility near any hazardous sites that might force an evacuation?

- Healthcare Facility: Although healthcare facilities can expose workers to some of the most dangerous hazards, the industry is also highly regulated. Agencies such as the Occupational Safety and Health Administration (OSHA) and the Joint Commission on Accreditation of Healthcare Organizations (JCAHO) speak directly to the unique hazards present in healthcare environments. These regulations will have bearing on your emergency planning and recovery activities. Questions to ask might include

- Would we be able to handle chemical or biological exposure?
- Do we have procedures for bloodborne pathogen exposure?
- How long can we operate without utilities?
- How would we handle an evacuation?
- How can we practice evacuations without impacting the patients?
- Does the facility have safe "areas of refuge" to which patients can be moved?

Of course, these are not the only questions you would ask when developing an emergency action plan. These questions represent just a sampling of the issues that must be addressed. The emergency planner must play the role of detective, looking for weaknesses in the current plan or its processes. It's okay to play devil's advocate. It's important to role play emergency scenarios, and don't be afraid to ask the hard questions. So far we've discussed the property, its geographic location, and the business it conducts. Usually these are issues that we can control. We should now change our focus to those external events to add to our risk.

Understanding Your Risk

The Risk Assessment Matrix

A logical starting point for determining risk would be to list any hazard that may occur. Upon completion of that list, you would then rank the hazards in order of probability. When completing this step of the process, it is advisable to consult with other sources that have already completed a risk assessment. For instance, many states and local agencies have already completed risk assessments and mitigation plans for your area. Public works departments, water management agencies, floodplain managers, and other agencies have done the legwork. Don't reinvent the wheel! If fact, many of the hazards that exist in your community may be "hidden" hazards which might be easily overlooked. To circumvent this possibility, consult with others.

> One emergency planner was smart enough to realize that a dam some 50 miles up river presented his facilities with additional risk. If you are not sure about the existence of dams in your area, contact the Association of State Dam Safety Officials (ASDSO). Their website is <http://crunch.tec.army.mil/nid/webpages/nid.cfm> (click on "State Links"). The majority of the dams in the United States have Emergency Actions Plans that identify areas that are likely to be flooded in the event of a breach.

Liability Versus Probability Versus Cost

After determining the probability of an event occurring, you should also determine the liability and the cost. A high-probability, low-liability event such as temporary loss of electrical utility may not warrant much consideration. However, a low-probability, high-liability event such as extended loss of utility must be planned for. A third component to consider is cost. Is the cost of the emergency event elimination or reduction worth the trouble? Is the return on investment worthwhile? For instance, you could construct a totally storm-proof, fire-proof, and secure building, but at what cost? Is that cost too high based on the liability or probability of an emergency event? These are the factors that must be considered when performing a risk analysis.

The frequency and duration of emergency events will also play a role in your planning efforts. We will discuss these factors later in the chapter, but realize that your geographic location will play a big role in the frequency and even the duration of an emergency event. This information should be taken into consideration when developing an emergency action plan.

The Federal Emergency Management Agency (FEMA) has created an excellent matrix for performing a risk analysis. (See Figure 4-1.) The matrix is effective because it allows you to assign a probability factor and human, property, and business impacts in addition to an internal and external resource scale. You would grade each of these components based on a scale of one to five. After grading out each emergency event, you simply add the number from left to right with the total tallied in the right hand column. The higher the number, the greater the concern. This matrix forces you to evaluate potential emergency events against a consistent measuring stick. It focuses your attention on the greatest needs and concerns.

If you think it's expensive to do emergency planning, consider the alternative. It is always more expensive to address the effects of an emergency after the event has happened. It is a fact: organizations that do not have an emergency action plan have a more difficult time recovering from an emergency event.

> Various reports have estimated that between 40 percent and 90 percent of businesses that experience disasters will go out of business within five years as a direct result of that disaster. Furthermore, 95 percent of those businesses were operating without the safety net that an emergency action plan provides.

Regarding liability, it would be a mistake to dwell too much on the "hundred year" flood and discount the less disastrous annual flood. In actuality, the annual flood may cause more damage over the long term because of its frequency. Take care not to focus too much on the "big" event at the expense of the "everyday" occurrences.

Type of Emergency	Probability	Human Impact	Property Impact	Business Impact	Internal Resource	External Resource	Total
	Hi Lo 5 ←→ 1	Hi Lo 5 ←→ 1	Hi Lo 5 ←→ 1	Hi Lo 5 ←→ 1	Weak 5 ←	Strong → 1	
Fires							
Severe weather							
Hazardous material spills							
Transportation accidents							
Earthquakes							
Terrorism							
Utility outages							
Other							

Figure 4-1. Risk Anaylsis Matrix

Online Hazard Maps and Databases

Wouldn't it be nice if you could retrieve historical information about natural hazard events in your area? You could review newspaper clippings and other announcements, but that process would be tedious and time consuming. You could ask people who have lived in the area for a long time, but memories fade and the details may be sketchy. Or you could simply visit a database that tracks hazards and allows visitors to perform a historical study of natural events and near misses. This database is a partnership between the Federal Emergency Management Agency (FEMA) and the Environmental Systems Research Institute (ESRI). Their goal is to provide multi-hazard maps and information to the public via the Internet. FEMA provides the historical hazard data and ESRI brings the information to life by providing GIS mapping and software. The end result is an invaluable tool for the emergency planner. The ESRI website can be found at <www.esri.com/hazards/>.

Understanding Your Responsibilities

OSHA Regulations

In addition to the moral responsibility you have to your employees and the fiscal responsibility you have to your employer, you have a distinct responsibility to OSHA. The Occupational Safety and Health Act requires that employers protect employees from fire and other emergencies. Part 1910 of the OSH Act addresses general industry, which covers most operating facilities. Specifically 1910.38 (a) (1-5) requires the following components:

- Emergency escape procedures
- Emergency escape routes
- Procedures to be followed by employees who address critical functions prior to evacuation
- Procedures to account for employees after evacuation has been completed
- Rescue and medical duties for those employees authorized to perform them
- The preferred means of reporting fires and other emergencies
- Names of persons who can provide further information about the plan.
- An alarm system that employees recognize for emergencies

In addition to these items, OSHA requires employers to train a sufficient number of people to assist in the safe and orderly emergency evacuation of employees. The employer is then required to review the plan with each employee covered by the plan at least once and then again whenever the employee's emergency evacuation duties are changed or when the plan itself is revised. The written plan must be available for all employees to review.

> The OSHA regulations referenced in the book are federal OSHA standards. More than 24 states have their own state-approved OSHA plans which supercede federal OSHA. OSHA requires state plans to be at least as stringent as the federal plan, and many are more demanding. To determine if your state has its own plan, visit the OSHA website at <www.osha.gov> and click on the link for state plans. If your state has its own plan, a link will be provided which will allow you to leave the federal site and go directly to your state OSHA website.

In addition to the emergency evacuation plan, OSHA requires a specific fire prevention plan. The requirements of the fire prevention plan are discussed in greater detail in Chapter 6.

It's important to remember that OSHA standards are considered to be minimum safety standards. OSHA regulations would supercede less demanding local codes that may exist; however, the opposite is often the case. Local codes can be more rigorous than the OSHA requirements, and when this occurs, the stricter local codes supercede the less strict OSHA regulations.

Signage

Safe exit from a building in the event of an emergency is not just a good idea, it's also the law. The Occupational Safety and Heath Act, the National Electric Code, the National

Fire Protection Association, and your local municipality all have specific requirements concerning building signage and emergency lighting. The purpose of these regulations is to ensure the safety of all building occupants as they exit your building. This includes not just employees but also visitors, vendors, clients, and anyone else who may be at your place of business. Many of these people are unfamiliar with your building and won't readily know the best route for evacuation.

OSHA has created a number of standards that address the need for proper signage. The OSHA regulations found at 29 CFR 1910.37(q)(1) - (8) include

- Exits shall be marked by a readily visible sign.
- Any door or passage that is not an exit but could be mistaken for an exit shall be identified by a sign reading "Not an Exit."
- No decorations or furnishings shall hide or obstruct the sign.
- Every exit sign shall be in a distinctive color and contrast with the interior finish.
- Every exit sign shall be illuminated by a reliable light source giving a value of not less than five footcandles.
- Every exit sign shall have letters not less than six inches high and three-quarters of an inch wide.

Although not an OSHA requirement, the Americans with Disabilities Act (ADA) requires building signage in many areas to be both visual (capable of being visually read) and tactile (capable of being physically read). Raised and Braille characters are required on signs that designate permanent spaces such as restrooms, exits, and rooms and floors designated by numbers or letters. ADA law does not specify that informational and directional signage be tactile, but it does specify other aspects such as the height and width of the lettering. Likewise, OSHA specifies certain letter sizes on exit signs. Remember, both ADA and OSHA are federal agencies and your local municipality or code enforcement board can create regulations that supercede these federal regulations if they are more strict.

> In many municipalities, low-level signage or floor-proximity exit signs are now required. These illuminated exit signs are usually installed six to eight inches above the floor level and are visible when ordinary exit signs installed near the ceiling level are obscured by smoke.

Material Safety Data Sheets

Ever wonder what's in the chemicals you use? Would you know what to do if you inhaled or ingested a strong chemical? What would happen if you attempted to wash an acid-based substance off with soap and water? Since a chemical spill or a release of hazardous gas might constitute an emergency, we'll spend some discussing hazard communication. The Occupational Safety and Health Agency has developed specific requirements concerning the chemicals you have in your buildings. OSHA's hazard communication standard, and specifically the "Right-to-Know" standard 1910.1200, requires that employers provide the following:

- Material Safety Data Sheets (MSDS)
- Chemical inventory list
- Written hazard communication plan
- Proper labeling on all containers
- Employee training

The first two items on the list, MSDS and chemical inventory list, play an important role in emergency planning and recovery and therefore should be addressed in the emergency action plan. The Material Safety Data Sheets (MSDS) are the standardized documents that provide this type of information. Information found on an MSDS includes

- Manufacturing company information
- Exposure limits
- Treatment if exposed to eyes
- Treatment if exposed to skin
- Treatment if inhaled
- Treatment if ingested
- First aid measures
- Storage requirements

The MSDS serves as a protection for employees who are over-exposed to a chemical or are exposed without proper personal protective equipment. If a current MSDS logbook is kept by the employer, an assisting employee could access the exposure information and act accordingly. The MSDS information could be given to rescue personnel when they arrive. Because of the thousands of substances in existence, they may not be familiar with the chemical that caused the problem. The MSDS will assist the rescue personnel in prescribing the proper treatment.

Another reason to have MSDS documents available is to protect emergency rescue personnel as they enter your building. Imagine your building is on fire and someone calls 911. The dispatcher sends the local fire department. As the firefighters arrive, they quickly survey the building and the fire. It's nearly midnight and there is no sign of employees in the building; however, there are a few automobiles in the parking lot. The fire doesn't appear to be too bad, so the firefighters enter the building to search for possibly trapped employees and to better extinguish the flames. Unbeknownst to the fire fighters, your company produces and stores large quantities of sulfuric acid and other hazardous chemicals. Additionally, a natural gas pipeline runs through the building. Just about the time the firefighters reach the core of your building, a series of large explosions occur creating a thick cloud of toxic gas. The lives of the firefighters were unnecessarily put at risk. Because detailed information was not available to the firefighters, they entered a building that contained multiple hazards. Many fire departments are requiring MSDS log books, floor plans, and other warning information to be present before they will enter a building.

How can you get copies of MSDSs for your logbook? MSDSs are available for all chemicals, even those not to be considered dangerous. By law, an MSDS must be provided to you by the manufacturer or the distributor of the chemical. Upon accepting delivery of a chemical, make sure the MSDS is included. If it is not, ask for it. When purchasing small quantities of a chemical from a retail store, ask a clerk for a copy. If they don't have a copy, you can visit the MSDS Search web site at <www.msdssearch.com/>. Simply provide some basic information and you'll be able to print out the needed MSDS.

Another piece of important information is the chemical inventory list (Figure 4-2). This document is required by OSHA and works in conjunction with MSDS information. The chemical inventory list identifies by name all the chemicals that are on site, the area in which they are stored, the date they were brought on site, and when they were removed (if applicable). The chemical inventory list would be especially useful for emergency rescue personnel because they could quickly glance at the chemical list and identify any concerns. Since the chemical inventory list identifies where the chemical is stored, the firefighter could cross reference this information against a floor plan and further gauge the risk of entry.

Chemical Name	Area Used	Date Introduced	Date Removed
Bleach	Janitor's closet	5/01/98	
Denatured Alcohol	Shop	11/15/00	11/20/01
Tile Mastic	Shop	11/15/00	
Shelia Shine	Janitor's closet	3/23/99	

Figure 4-2. Example of a Chemical Inventory List

Americans with Disabilities Act

The Americans with Disabilities Act (ADA) offers many excellent suggestions for assisting those with disabilities during evacuation of a building. This issue is more difficult than most think. If handled improperly, the disabled can be hurt and the evacuation of other building occupants can be delayed. With this in mind, special consideration should be given to persons with disabilities. Hearing, vision, mental, and mobility issues need to be addressed prior to an emergency event, not during it.

In many cases, it is very easy to observe certain disabilities and make allowances for them. However, persons with disabilities, especially disabilities that are not readily visible, are often hesitant to reveal their disabilities. This presents the emergency planner with a very difficult task. How can the planner develop steps to protect the employee if the planner is not aware of a limitation? To bridge this gap, the planner should provide all employees the forum to privately reveal any limitations that may affect their ability to evacuate the building. Employers do have the legal right to ask employees if they have a disability that would hinder them from evacuating the building. To avoid drawing attention to a disabled employee, all employees should be asked this question, not just employees who have an obvious disability. A great way to address these issues without making disabled employees feel uncomfortable is to get them involved in the evacuation planning effort. They will likely be able to add insight that most employees would have been unaware of.

> The Americans with Disability Act defines a disabled person as anyone who has a physical or mental impairment that substantially limits one or more major life activities. This may include vision, hearing, mobility, mental, speech, or other serious health impairments including diabetes, heart disease, cancer, asthma, and alcoholism or drug use.

Don't forget about temporary or non-protected disabilities that may affect the employee's ability to safely evacuate the building. These issues may include pregnancy, knee or hip problems, broken bones, obesity, and asthma. Pregnancy is not a protected ADA classification; however, a pregnant women attempting to walk down 70 flights of stairs during an evacuation may have difficulties. The intelligent emergency planner understands that these temporary limitations must be addressed and will establish procedures to assist this person. Another example is the person with allergies or asthma. Under normal working circumstances, the symptoms are not revealed. Even if that person were required to evacuate the building and walk down 70 flights of stairs, the physical condition would not manifest itself. However, during the stress and excitement of an actual emergency coupled with the presence of smoke, the physical issue becomes a major liability. These are the types of non-protected or temporary disabilities that need to be addressed.

Partner System

When evacuating a building there will inevitably be those people who require assistance. Many organizations have established partner systems (sometimes known as buddy systems) and have been successfully using them. Partner systems are not perfect, though. They work when both parties are aware of and respect the arrangement. Partner systems fail for a number of reasons including

> Prior to offering assistance to a disabled person, ask them how you can help them. Don't be pushy! Ask how they can best be assisted or moved and whether there are any special considerations you should know about. Also ask them if they have any items that need to come with them. Show this person respect and do not treat them like a child.

- The assisting partner is absent from the workplace
- The assisting partner is in the building but at a different location
- The disabled partner is absent from the building but the assisting partner does not realize it
- The disabled partner is in a different area of the building
- The assisting partner has not been trained on how to react and assist
- The assisting partner is not physically appropriate for the task
- The assisting partner or the disabled partner are not patient with one another
- The assisting partner becomes frightened and leaves without the disabled partner
- The disabled partner evacuates without the assisting partner or with the help of someone else

These examples may seem like non-issues. Certainly we wouldn't leave someone who was depending on us, right? In times of emergency, we sometimes do strange things. The partnership must be one that is mutually respected and understood. Training must be delivered to both parties on a regular basis and prior to an emergency event. Some of the possible solutions to the problems inherent to partner systems include

- The disabled partner should play an active role in selecting their partner
- The assisting partner should understand the potential rigors of evacuating a disabled person
- A system should be established to communicate absences or "not founds" to floor captains
- A system should be established for disabled partners who do not use their assisting partners (and safely exit the building) to communicate their status to the exterior group leader
- A backup partner should be assigned and trained for each disabled partner
- The assisting partner should work near the disabled partner

Partner systems work. If both parties understand the assignment and take it seriously, lives can be saved and evacuations can be conducted more efficiently.

Visual Impairments

We sometimes fail to comprehend the importance of our senses, in particular our vision. Imagine performing basic life functions without eyesight. Imagine not knowing what obstacles are in your path. Imagine not knowing where the closest emergency exit is located. For obvious reasons, those with visual limitations require special attention during an evacuation. By using a partner system, the visually impaired can be led to safety. Because many of us are not accustomed to assisting the blind, here are a few reminders:

- Announce your presence as you enter the work area.
- Do not shout at them; they are blind, not deaf.
- Do not grasp a visually impaired person's arm.
- Ask if they would like to hold onto your arm as you exit.
- Ask them to explain what assistance they need.
- Before walking towards the exit, let them know that debris or crowds are present.
- Verbally provide warning of stairs, ramps, doorways, or other obstacles.

Once outside the building, do not abandon the visually impaired person. He or she will be in unfamiliar surroundings, in a crowd of moving people, and perhaps in the way of emergency rescue personnel. Bring this person to a staging area or other safe haven.

Hearing Impairments

Census reports identify over 22 million people with serious hearing impairments, making this the largest class of people protected by ADA law. Of these 22 million people, nearly two million are entirely deaf. Given these large numbers, there is a good chance that you will have people in your building with hearing impairments. Again, we often take our senses for granted. Imagine an alarm being sounded and an evacuation being ordered over the loudspeaker. Without hesitation, we would make our way toward the exit routes and out of the building. Now think about people with hearing impairments. Let's assume they are working at their computers, fully engrossed in what they are doing. Based on their hearing limitation and their involvement in their work, they would be oblivious to the emergency event. Like those with visual limitations, those with hearing limitations also require special attention during an evacuation. The partner system is the best way to assist those with disabilities, but because many of us are not accustomed to assisting the hearing impaired, here are a few reminders:

- Speak directly to the individual, not to their interpreter.
- Get their attention by touching them and keep it through eye contact.
- When entering their office, flick the lights to get their attention.
- Clearly state the problem using gestures and pointing.
- Write a brief statement if the person does not seem to understand.
- Be patient, especially if they do not comprehend what you are saying.
- Give them a flashlight so they can read lips if the lights go out.

In many organizations, those with hearing disabilities are given a pager that vibrates in the event of an emergency. The alphanumeric pagers work best because a text message can be sent to give the employee an idea of what is going on and how they should proceed. Audible alarms should also be coupled with blinking strobes that would alert an employee who could not hear. In some hotels, persons with hearing disabilities are given rooms with beds that vibrate if the fire alarm is activated. The tools are out there. There is no good reason for not accounting for the unique needs of those with hearing disabilities.

Summary

For those organizations without an emergency action plan, the task of creating one can seem insurmountable. Like any large task, you must break the development of your plan into small manageable pieces. A good starting point is to perform a risk analysis. This analysis takes into consideration your geographic location, your resources, your core business functions, and the emergency events most likely to affect your business.

Of course, the development of a good emergency action plan doesn't happen without an investment of time and money. You may have to convince management of the importance of your endeavors. This should not be too difficult given the historical accounts of businesses that were unable to survive an emergency event. Much like an insurance policy, you hope you'll never need to use the emergency action plan, but that doesn't mean you can go without one.

5

Creating the Emergency Action Plan

The Emergency Action Plan

Creation of the emergency action plan is not a simple process. Risks must be evaluated. Critical business functions must understood. Liability of an emergency must be compared to probability of it happening. Contingency plans must be developed. Assignments must be given and employees must be trained. At some point, all of this information must be written down in some type of logical format. The compilation of this information becomes your plan. Unfortunately, the plan is often viewed as unnecessary and even unrealistic. The only reason it exists is to satisfy management's desire for something tangible. They need proof that the investment they made in emergency planning was worth it. In an effort to satisfy their concerns, we equate quantity with quality and we produce a big thick emergency planning manual!

In a perfect world, the emergency action plan would provide valuable, up-to-date information that could quickly be referenced during an emergency event. It would not be so large that it hides the important stuff. It would not contain unnecessary information. Nor would the plan be too small or too vague. The emergency action plan is not static, it is a living document. It will change and evolve over time. The initial time spent developing the plan will not likely need to be repeated, but the document must be maintained. The initial development of the document may be painful and time consuming, but it is project well worth the time and expense.

This book was created to assist emergency planners with the development of such a plan. It's impossible to provide an emergency action plan that satisfies the needs of all organizations. A turnkey plan that can be purchased and simply placed on a shelf until an emergency occurs does not exist. Even with the help of this resource, you will be required to research and develop a plan that is specific to your business. With this in mind, let's discuss the plan, its purpose, and what is does in greater detail.

What Does It Do?

The emergency action plan exists to save lives and reduce injuries in the event of an emergency. It provides the means and methods to help you plan, survive, and recover from an emergency. Because lives, property, and business continuity are at stake, the emergency action plan must be considered an important part of your organization's business plan. It must be championed at the highest levels of the organization. It should educate employees and serve as a basis for employee training. Because people and processes do change, the emergency action plan must be reviewed on a regular basis and updated as required.

What Does It Contain?

Quite simply, your emergency action plan should contain all pertinent information that will allow you to plan, survive, and recover from an emergency. This will require an evaluation of your building, type of business, and geographic location. The plan must contain names and phone numbers of employees who are able to offer assistance in the event of an emergency. In 29 CFR 1910.38(a)-(b), OSHA has developed regulations that identify the minimum requirements of an emergency action plan:

- Emergency escape procedures
- Description of potential hazards
- Emergency escape routes
- Procedures to be followed by employees who remain to operate or shut down critical building operations before they evacuate
- Procedures to account for all employees after evacuation has been completed
- Rescue and medical duties for those employees assigned to perform them
- The preferred means of reporting fires and other emergencies
- Names or job titles of employees who can further explain the emergency action plan

OSHA also requires that each facility have a fire prevention plan in place. This plan is specifically tailored for the threat of fire. The employer is required to train employees in aspects of the plan that pertain to their exposure. Components of the plan include

- A list of fire hazards that exist in the workplace
- Names or titles of employees who are responsible for fire source hazards
- Names or titles of employees who are responsible for fire alarms and fire fighting equipment
- Housekeeping procedures that limit fuel sources
- Employee training techniques and frequencies

In addition to these minimum OSHA requirements, a good emergency action plan would include the following items:

- Statement of policy, also known as a mission statement

- Floor plans with exits, fire extinguishers, pull stations, first aid stations, and other important features marked
- Location of MSDS information
- Names and numbers of external emergency service providers
- Inventory lists of emergency supplies
- Maintenance schedules for generators, fire sprinkler, alarm systems, and other related emergency equipment
- Chain of command established for emergency events

Who Gets a Copy?

Who gets a copy of the emergency action plan? That depends largely on your organization's preference. The actual emergency action plan can be quite large and detailed; if given in written format, most employees will not read it. OSHA does not require every employee to have a copy of the plan. Instead, OSHA requires that the plan be kept at the workplace and made available to all employees. A suggested best practice would be to provide all members of your emergency management team with a completed copy of the plan and simply make the plan available to all employees. Many organizations have done this by putting the plan on the company's intranet website. Other organizations have provided employees with a scaled-down version of the emergency action plan, usually during new employee orientation. A summary version explains proper evacuation procedures, alarms, staging areas, and other basic issues.

One important issue is the regular revision of the emergency action plan. You do not go through all the effort of creating a plan and then let it sit on a shelf. The plan should be reviewed at least annually and revised as needed. When changes are made to the plan, training should be provided to all employees affected by the changes.

In the event of an emergency, the action plan will be a vital reference. You will consult the plan on a regular basis, looking for names and number of key support personnel. You'll use the plan as a guide for making decisions. The plan will ensure that all the necessary actions are taken to promote business continuity during and after the emergency. Bottom line—the plan is of utmost importance! But think for a moment about the actual plan. Where do you keep it? Probably in your office. Would you be able to access the plan if the building burned down? What if an earthquake occurred and the building was unsafe to enter? What if everybody on the emergency management team kept their plans in their offices? Of course, you could go online to the company's intranet site and print it out. What if the electricity was out for an extended period of time? What if your computer was damaged during the emergency event? For these reasons, it is imperative that you keep a copy of the emergency plan at an offsite location such as your vehicle or home. It makes sense to keep a copy on your laptop as well. Don't be caught in a situation where the plan is not available!

The Emergency Management Team

Every good plan is supported by a team of dedicated people. The plan is just a reference manual. Decisions still need to be made. Direction must be given. Human interaction and leadership is a necessity. The plan comes to life when somebody picks it up and actually uses it. Furthermore, the emergency management function is not a one-person job. Instead, a team of people will be involved, each with a specific purpose.

Creation of an emergency management team is essential early in the planning process. Expertise will be needed from all areas of the organization. If the emergency management team consists of employees with similar backgrounds, education, responsibilities, and job titles, the plan will be written in a vacuum. Valuable expertise from outside your department will not be tapped into, so make sure you include a diverse group of employees on your team. This team will likely grow as the plan becomes a reality. Floor captains, runners, evacuation partners, and command post leaders will be need to be elected. For now, though, the team should be kept small but functional.

The length of time a member would sit on an emergency team is subjective. Some organizations like to rotate employees in and out of the team on an annual basis. The belief is that by allowing a greater number of employees the chance to sit on the team, a greater number of employees will respect the plan and be able to react during crisis. It's also believed that in the event a team member leaves the organization, the vacancy can be quickly filled by a past member. This may not be a bad idea, but it does have its problems. If the expertise of the team leaves each year and untested, and perhaps unmotivated, employees take over, you'll lose some of the continuity of the team. To compensate for this, instead of rotating the entire team, try rotating one or two positions each year. This way, the bulk of the emergency team is left intact while employees with fresh ideas and viewpoints can be introduced.

> Because emergencies can happen anytime during the day or night, the emergency management team may need to respond after hours. A phone list of current members and responsibilities should be made available to each member of the team. For ease of use, print the information on a credit card–sized paper and laminate it. Team members will be able to keep this list in their purses or wallets and access it quickly in the event a response is needed.

Identifying Team Members and Responsibilities

When developing your emergency management team, be sure to include a diverse roster of employees. In addition to the obvious goal of saving lives and reducing injuries, your goal is to identify critical business functions that may be interrupted as a result of an emergency event. This is not usually accomplished through the eyes of a single individual. Once the diverse team of subject matter experts has been finalized, assignments and a hierarchy should be developed. During a crisis there must be a clear line of responsibility and command. Listed below are samples of some of the expertise needed on your team:

- Risk Management Specialist
- Logistics and Procurement Manager
- Media Relations Contact
- Human Resources Personnel
- Facilities Maintenance Manager
- Legal Counsel

Listed below are a few of the assignments that should be given:

- Emergency Director
- Business Continuity Manager
- Command Post Leader
- Group Leader
- Floor Captains
- Fire Brigade
- Evacuation Assistants
- Runners

For a more in depth explanation of these assignments, refer to pages 157–158. [chapter 6 after "Accounting for Employees and Visitors"]

Internal and External Support

The risk analysis matrix discussed earlier includes a column for internal support and a column for external support. Although you'll want to take advantage of your internal resources and employees, that may not always be possible. Key to surviving any emergency event will be the support of external resources and vendors. Early identification of these resources is essential, as is involving them in the planning process.

When identifying internal and external resources, consider to how quickly the resource will be available. This plays an important role when discussing internal resources. You may have an excellent staff of highly trained maintenance technicians who are skilled at solving building-related problems, but the emergency event may cause them to be unavailable. In the event of labor unrest, your internal resources may not be willing to cross a picket line. When an emergency event occurs, your employees and their families may be directly affected and therefore unavailable, so make plans for a backup.

Insurance providers play a critical role in your emergency planning and recovery efforts. Unfortunately, we often have a very fuzzy idea of what coverage we have in place. The time to educate yourself on these important issues is prior to an emergency, not after it. Many organizations believed they were covered by their insurance policies, only to find later that an amendment to the policies excluded a specific event. Since coverage terms, property valuation, property size, and other issues change on a regular basis, you should meet annually with your insurance advisor to discuss your policy. Typical questions might include the following

- How will the property be valued (replacement/actual/other)?
- Am I covered for spoilage of goods?
- Do I have all the policies needed (wind/flood/storm/contents/liability)?
- Do I have a policy protecting me in the event of lost income?
- In what way does the provider require proof of loss?
- What additional loss documentation is needed and is it stored in a safe place?
- If I do have a lost income policy, and what is the duration?
- Does it make sense to self-insure?
- What are my deductibles and are they at proper levels?

> On October 30, 2000, the president signed the Disaster Mitigation Act of 2000 (Public Law 106-390). This new law encourages and even rewards local and state disaster planning. It serves as a bridge between numerous local agencies and the state, thereby increasing timely communication between these groups. Those organizations that are not considered public entities are encouraged to contact their local and state agencies to ensure compatible planning.

Identifying Vendors and Contractors

Contractors come and go; it is, therefore, imperative that you review your list of contractors with active purchase orders annually at a minimum. Be certain that you have updated office and cell phone numbers. In heavily populated metropolitan areas, area codes tend to change on a regular basis, so make sure you have the latest area code as well. An excellent form for tracking contractor availability is depicted in Figure 2-6. [p. 36, Chapter 2]

Identifying a good contractor is just a starting point. A contractor may agree to offer assistance after the emergency event, but because of circumstances out of their control, they may not be available. Access roads could be blocked, their office could be heavily damaged, or some other logistical problem might exist. For this reason, you need to have alternate contractors available. The Contractor Availability Form includes space to list alternate contractors.

Using the services of a contractor after an emergency event sounds like an easy process. Simply call the contractor, issue a work ticket, tell them where to show up and what to do. In actuality, the process is not so simple. Often you know the contractor's services are needed, but you're not sure of the scope of work. For instance, after a tornado, you contact your landscaping contractor and dispatch them to your corporate headquarters. Based on the reports you've heard, trees are down and blocking access roads and parking lots. Other trees are not down but pose a danger and require removal. There will probably be some irrigation repairs as well. Because the scope is unclear, you simply instruct the contractor to get the property back in order. Ambiguous requests like this are likely to cause over-billing. To account for these issues, meticulous records must be kept. Depending on the relationship you have with the contractor, you may require a field inspector from your organization to give approval prior to commencing work. This may keep costs under control, but it really slows down the recovery process. A better way to handle large projects that are discovered by the contractor in the field is to require them to document the damage with a camera or video camera prior to commencing the cleanup efforts. If your relationship is good with the contractor, this process will protect you and keep the contractor busy. You may also elect to place a cap on cleanup efforts. For instance, any job that exceeds $2,500 must first be approved by the facilities department. By blending a quick efficient cleanup with cost protection measures, you'll get the job done without a financial hangover the next morning.

Prior to using the services of a contractor after an emergency event, you must iron out a few of the contractual details. The emergency creates unique circumstances. The contractor will bring in extra help. Many of the contractors will be providing assistance around the clock. Specialized equipment may need to be rented. When you consider the possible issues, you'll begin to realize the importance of creating contractor specifications for emergency events. Examples of these issues include:

- Overtime rates (time and a half, double time)
- Hotel reimbursement
- Mileage reimbursement
- Expense and receipt requirements
- Food reimbursement
- Laundry reimbursement
- Equipment rental responsibilities
- Personal protection equipment requirements
- Additional insurance requirements

Because much of the help you use will not be your primary contractors, you'll be tapping into companies that may not meet your usual company expectations. Specifically, insurance requirements come to mind. Do you feel comfortable contracting with companies that carry inferior amounts of liability insurance? It's a risk and you must weigh that risk against the restoration needs you face. Be very careful with this decision. Remember, after an emergency unusual and dangerous conditions will likely exist. If this contractor does not have adequate insurance for normal working conditions, how can you feel comfortable using them in a dangerous environment? Instead of lowering your standards, spend more time locating quality alternate contractors who already have the required coverage in place.

The indemnity agreement should be a standard document signed by all contractors, regardless of their insurance amounts. This document protects your company from lawsuits and judgments as a result of the contracted assignment. As a last resort, you may elect to use contractors with inferior coverage amounts if they agree to sign an indemnity agreement. This document does not provide total protection, but at least it is a start. Of course, when signing legal contracts, always consult with your legal counsel beforehand. An example of an indemnity agreement coupled with a safety clause may look like this:

> Contractor hereby agrees to indemnify and hold harmless [company name] its agents, officers, successors, and assigns against any and all claims, suits, judgments, and damages of any kind resulting from this contract or the work supplied by the contractor, its employees, or subcontractors.
>
> Furthermore, contractor shall be solely responsible for supervising its employees, providing personal protection equipment, training its employees, record keeping, and adhering to all federal Occupational Safety and Health Act regulations or state equivalent.

Stocking Emergency Supplies

Figure 5-1 identifies items that should be stored as soon as storm season begins. The inventory list also has columns for multiple locations. Of course the actual inventory and the numbers needed depend on your individual location.

Items that have a limited shelf life such as food, batteries, and fuel need to be rotated out of storm inventory and into common inventory. This best occurs at the end of storm season. For instance, if you have stored three dozen AAA batteries for your radios and you have not used them at the end of storm season, use them elsewhere. Don't make the mistake of sitting on those batteries for five years and find that they have expired when you really need them. Don't make the inventory more complicated than it needs to be, but remember the golden rule of inventory management: First In = First Out.

In preparation for emergency events, it may be necessary to store a limited amount of food supplies. This can present a problem because the shelf life of most foods is very limited. Other foods require preparation including cooking. A popular solution to this dilemma is the use of MREs (Meals ready to eat). MREs are fully cooked meals originally developed for the military. Depending on storage temperatures, the shelf life of an MRE can exceed ten years without refrigeration. The food is not freeze dried, and each item is cooked and sealed in individual high strength pouches. Full meal MREs typically

STORM PREPARATIONS
Inventory of Materials

	Location A	Location B	Location C	Location D
TOOLS / EQUIPMENT				
Chain saws (18') with 1 gallon gas can	1	1	2	1
1 quart bar & chain oil	2	2	4	2
2-cycle engine oil	2	2	4	2
Shovels	4	4	4	4
Wet vacs (25 gallon)	2	2	2	2
Fire estinguisher	4	4	6	4
Axe	1	1	2	1
Crowbar	1	1	1	1
Push Brooms	2	2	2	2
Chains and locks	2	2	2	3
Extension cords	3	3	4	3
Circular Saw	1	1	1	1
Screw guns / drills	2	2	2	2
Duct Tape	3	3	3	3
60 mil plastic liner (roll)	2	2	2	2
Orange spray paint	1	1	1	1
Ladder	1	1	1	1
Fans	2	2	4	2
Barricades	4	4	8	4
Emergency caution tape	2	2	2	2
FUEL SUPPLIES				
Gas can (5 gallon)				
Propane gas (Bottles for lanterns)				
Propane - 20lb Bottle for gas grill				
LIGHTING				
Propane lanterns				
6-volt lanterns				
Flashlights				
Portable quartz lighting				
BEDDING				
Folding cots				
Wool blankets				
CLOTHING				
Rubber boots, Size 10				
Rubber boots, Size 11				
Rubber boots, Size 12				
Work gloves				
Rain jackets, large				
Rain hoods				
Rain pants				
Hard hats				
Dust masks				
Respirators				
Safety glasses/goggles				
Reflective vests				
ROPE/WIRE				
Tie wire (50' Roll)				
Rope, 1/2" (300' Roll)				
Rope, 3/8" (200' Roll)				
PUMPS				
Sump pump w/hose and discharge hose				
MEDICAL				
First Aid kits				
Aspirin				
Sunblock				
Insect repellent				
WATER HOSES				
2 1/2" 6' Hose pkgs.				
SAND SUPPLIES				
Burlap bags				
Sand delivered (by load)				
GENERATORS				
Portable generator				
Fuel (unleadead gasoline)				
Oil (for Diesel Generator)				

Figure 5-1. Storm Preparation Inventory List

STORM PREPARATIONS
Inventory of Materials

	Location A	Location B	Location C	Location D
RADIOS / BATTERIES				
Radios, NOAA				
Radios, 2-Way				
Clock (battery operated)				
Batteries - (size AA to D and 6V lantern)				
Calculator				
WATER / ICE / FOOD				
Ice chests (50 qt)				
Ice				
Meals Ready To Eat (MRE's)				
Matches				
Utinsils				
Garbage bags				
Water purifying tablets				
Bottled water - 5 gal. bottles				
Can opener (manual)				
BUILDING MATERIALS				
Plywood (4' x 8' x 5/8" sheets)				
2"x4"x 8' studs				
Screws				
Nails				
RENTAL EQUIPMENT				
Port-O-Potties				
Boat w/oars and accessories				
PHOTO EQUIPMENT				
Digital camera				
Disposable camera (Waterproof)				
Video				
CASH				
Emergency cash				
OFFICE SUPPLIES				
Steno pads				
Highlighters				
Pens				
Pencils				
Clipboards				
Paper clips				
Staples				

Figure 5-1. Storm Preparation Inventory List, continued

include a main entree, a side dish, a desert, crackers and spread, beverage base, a spoon, and an accessory packet. Although cooking of MREs is not required, some foods taste better if heated. Chemical MRE heaters are available that will effectively heat your food in a matter of minutes using only a little water.

A number of organizations produce and package MREs. You do not need to be a member of the military or other government agency to purchase them; in fact, many personal households keep MREs in stock in the event of an emergency or for family camping trips. MRE packages can be purchased that also include lights, batteries, personal hygiene products, and first aid supplies.

For an excellent source of a full range of MRE products, contact

Epicenter
384 Wallis #2
Eugene, OR 97402
(541) 684-0717
<www.geoduck.com/epicenter>

Blueprints and As-Builts

A number of tools can assist emergency teams during search and rescue missions after an emergency event. Most of these tools are readily available to the planner, such as floor plans annotated with exits, fire extinguishers, and first aid kits. Other tools such as the Urban Search and Rescue Grid are created by highly specialized professionals and are particularly important for use in multi-tenant high-rise buildings. When creating your emergency plan, start by acquiring an updated copy of the building floor plan. In a leased building you will need to speak with the property owner or its agents. If the building is company owned, you should speak to the facility management department. Over the years buildings are renovated, so the floor plan that you have may not be an accurate representation of the building's configuration. Take time to ensure the accuracy of the floor plan. If you cannot locate any floor plan information, or if the plan is hopelessly outdated, it would be advisable to create a basic revised floor plan. If you have the tools and expertise in-house, use them to create a basic floor plan. It's not necessary to go to into great detail on your floor plans; remember, the purpose is not to build the structure. The purpose of this floor plan is to identify windows, doors, emergency exits, alarm panels, fire pull stations, and other emergency-related items. A basic single line floor plan minus the dimensions, details, and other non-essential items will be sufficient. When developing an emergency floor plan, remember to include important details such as

- Drainage
  - Storm drains
  - Roof drains
- Egress Concerns
  - Designated escape routes
  - Stairways
  - Windows
  - Doors
  - Emergency exits
  - Evacuation staging areas
- Fire Safety
  - Fire alarm pull stations

> Take pictures or a video of the facility and its assets before and after an emergency event. This documentation will prove beneficial when dealing with insurance adjusters.

- Fire extinguishers
- Smoke detectors
- Water sprinkler testing valves
- Alarm panels
- First Aid
  - First aid kits and stations
  - Automated External Defibrillators (AED)
  - Eye wash stations
- Specialty Areas
  - Mechanical rooms
  - Telecommunications rooms
  - HazMat storage locations
  - MSDS locations
  - Restricted areas
- Utilities
  - Gas mains and valves
  - Utility shutoffs
  - Water lines and valves
  - Sewer lines
- Power Systems
  - Generator and switchgear locations
  - Generator powered circuits
  - Uninterruptible power system location

Critical Building Information

Critical building information must be made available during and after emergency situations. This is usually not a problem if company employees are available. Facility management or maintenance personnel are a ready source of critical building information, but what would happen if the emergency event occurred after business hours? For instance, what if a fire were to start in your building after hours? When the fire department arrives, will they have keys to your building? How will they locate a floor plan? Will they be able to locate your MSDS information? If the fire department does not have keys, they will enter your building forcefully, but considering the emergency, that will be the least of your problems. The floor plan would be helpful but, again, not necessary. The MSDS information, however, is very important to the firefighters. Without an idea of what chemicals are being stored on the premises, the firefighters are putting themselves in a very dangerous situation. It would be unfair to ask any rescue personnel to enter your building without first identifying the hidden hazards.

The KNOX-BOX® Rapid Entry System provides a means for non-destructive emergency access to commercial and residential property. Key vaults are available in small sizes just large enough to hold a set of building keys. Larger vaults are available that can hold keys, MSDSs, floor plans, and other critical building information. These vaults are specially keyed to the requirements of your local rescue personnel, much like an elevator key. Check with your local fire department to see if KNOX-BOX systems are used in your area.

Fire Extinguishers and Means of Egress

Portable fire extinguishers provide the first line of defense when fighting a small fire. In many cases, quick response to a fire coupled with the proper extinguisher can prevent loss of life, damage to property, and reduced productivity. In fact, if the fire is contained and extinguished, an evacuation of building's occupants may not be necessary. At worst, a partial evacuation of those areas nearest to the fire may be all that is required.

Portable fire extinguishers are critical for extinguishing small fires in their early stages. Prior to using a portable fire extinguisher, the operator should be properly trained. Three major of classes of fire exist, and if the emergency responder chooses the wrong fire extinguisher for a specific fire, he risks causing further harm. Proper training starts with identification of the fire source and proper selection of extinguisher. The three major classifications of fires are

- Class A Fires: Fueled by common combustible materials such as wood, cloth, paper, and many plastics. Use a water-based extinguisher with a green label and Class "A" identifier.
- Class B Fires: Fueled by flammable liquids and gasses such as gasoline, kitchen grease, and paint. Use a carbon dioxide or aqueous film forming foam (AFFF) extinguisher with a red label and Class "B" identifier.
- Class C Fires: Fueled by live electrical components or circuits including appliances, tools, and wires. Use an extinguisher with a blue label and Class "C" identifier.

Fire Suppression Systems

In the event of a fire, quick response from local firefighting professionals is critical. In reality, the response is often not quick enough. The firefighters often arrive in time to extinguish the fire, aid the victims, and ensure that the fire does not spread to adjacent properties. Unfortunately, the damage to the subject property is often great. A sensible solution, especially in buildings that are unoccupied during the evening, is the installation of fire sprinklers. In many areas, fire sprinkler systems are mandatory on new construction and renovations. In fact, The National Fire Protection Association (NFPA) has been a leader in developing standards for the fire protection industry. The NFPA has determined that installation of a fire sprinkler system coupled with smoke detectors could reduce deaths attributed to fire by 82 percent. Armed with this information, we can appreciate the importance of functional fire suppression systems.

There are a number of fire sprinkler options:
- Dry Pipe systems are used in areas where freezing conditions exist. Pressurized air is maintained in the sprinkler pipes. When a fire occurs, the heat ruptures a seal in the sprinkler head, the air escapes the pipes, and the water exits the sprinkler heads.
- Wet pipe systems do not use pressurized air. When a fire occurs, the heat ruptures a seal in the sprinkler head and the water exits the sprinkler heads.
- Deluge systems are similar to dry pipe systems in that they use pressurized air in the pipes. This system may be activated manually or automatically.

> In the event of a fire, you would expect the fire sprinkling system to work properly, but will it? If you have not completed regular testing and maintenance on your system, you won't know if the system will work properly until a real fire tests it. That's not a good option! Regular testing of your fire sprinkler systems is critical to an effective emergency plan. The National Fire Protection Association has created standards that define testing requirements for fire sprinkler systems. The NFPA 25- Standard for the Inspection, Testing, and Maintenance of Water-Based Fire Protection Systems standard suggests opening the sprinkler systems test valve on a quarterly basis to simulate an actual fire event.

After the terrorist attacks of September 11, 2001, engineers determined that the World Trade Center Towers collapsed due to weakened steel members and not due to the violent collision of the two aircraft into the buildings. The aircraft were fully loaded with fuel, and as the fuel burned inside of the towers, temperatures reached in excess of 2,000 degrees. The World Trade Center Towers contained fire sprinkler systems, but the impact of the aircraft ruptured the lines and rendered them useless. This should not lessen the credibility of fire sprinklers as an effective way to extinguish fires, especially considering the unusual nature of this particular emergency.

On that same day, another terrorist attack occurred at the Pentagon. An airplane taken over by terrorists slammed into wedge one and wedge two of the Pentagon, penetrating four out of the five rings that make up the structure. Engineers noted that damage to the newly renovated wedge one was minimized thanks to the fire sprinkler system. The fire did not extend to areas beyond where the fuel was spilled. At wedge two however, fire spread easily and rapidly because a fire sprinkler system had never been installed in this soon-to-be renovated area.

Pre-Planning and the Urban Search and Rescue Grid

When the fire department responds to a building after an emergency, critical building information is needed if firefighters are to effectively and safely deploy their resources. Consider the fact that most rescue personnel will be unfamiliar with your building, its floor plan, its contents, and other important building issues.

To assist fire rescue personnel with this dilemma, former fire officer Curtis Massey founded Massey Enterprises, Inc., in 1986. Since that time Massey Disaster Plans have become an industry standard. These plans are taught as part of the curriculum at the National Fire Academy and are also the only pre-planning system supported by every major fire department in North America. These plans address the emergency from the viewpoint of rescue personnel and include the not-so-obvious features that are critical to rescue personnel. Fire rescue, the Federal Emergency Management Agency, police departments, and other emergency personnel are assisted by the Massey Disaster Plans. These plans are invaluable for all types of emergencies including fires, gas leaks, hazardous materials incidents, terrorism, biological/chemical attacks, mass shootings, hostage situations, technical rescues, and natural disasters.

Armed with the critical building information on the plans, fire rescue personnel can best utilize the building systems to their greatest advantage. Detailed, easy-to-read graphics complement the text. Floor plans, site plans, and riser diagrams show the fire department where everything is located—both horizontally and vertically.

The Massey Disaster Plans include two distinct features—the pre-plan and the urban search and rescue grid. The pre-plan features color-coded graphics that fire fighters understand and are able to quickly decipher. Items such as HVAC, shut-off valves, communications, alarms, and water supplies are noted on the pre-plan.

The second feature of the Massey Disaster Plan is the Urban Search and Rescue Grid. This document identifies support columns and other major features of the building that would enable firefighters to systematically search the building even if it experienced a collapse.

Because of the detailed nature of these plans, their completion make take months and cost thousands of dollars; however, the effort is well worth the expense, especially if you manage a high-rise building. Massey Enterprises, Inc., can be reached at

Massey Enterprises, Inc.
3300 Building, Suite 400
397 Little Neck Road
Virginia Beach, VA 23452
Phone: (757) 340-7800
Fax: (757) 340-4510
<www.disasterplanning.com>

Alarm Systems

OSHA requires employers to devise a method for notifying employees of an emergency. These emergencies might include fire, chemical spill, bomb threat, or other related events. The alarm must be able to be heard above the noise level in the workplace. In some areas—shop or manufacturing plants, for instance—this may be difficult to do. In these cases it is recommended that a secondary alarm warning system be used. For instance, an audible alarm coupled with a visual alarm such as a flashing strobe would provide much better protection. The employer must also explain to the employees the preferred means of reporting emergencies. Methods may include manual pull box, public address systems, radio, or telephone.

The Americans with Disabilities Accessibility Guidelines (ADAAG, 4.1.3 and 4.28) outline specifications for emergency alarms to ensure that they are accessible to persons with disabilities. ADAAG specifications for visual alarm devices specify intensity, flash rate, mounting location, and other characteristics. One major difference between the installation of audible alarms and visual alarms concerns location. It is always a requirement to install audible alarms less frequently than visual alarms. This requirement makes good sense because the reach of an audible alarm is much greater than a visual one. A visual alarm can be blocked by line of sight and unseen if the individual is in a room that does not have a visual alarm.

Understanding Your System

An alarm system is an integral part of your property's overall emergency management plan. The system must be operational and available during all hours of the day and night. It must be kept maintained, and you must test it on a regular basis. Of course, most organizations will contract this work out to others because they don't have the expertise in-house. If you elect to do that, it's not a problem; however, it would be advantageous to develop an in-depth understanding of your alarm system.

Although this book does not pretend to be a technical resource that provides an overview of alarm and fire suppression systems, we will discuss a few of the basic regulations and guidelines concerning your alarm systems.

The first step is to familiarize yourself with the alarm system currently employed at your property. Start by identifying the systems in place. Reference your manufacture's literature including the technical information. If you are subcontracting maintenance or repairs to an alarm company, tap into their expertise. Ask them questions. Walk with them as they perform the regularly scheduled maintenance functions. In fact, invite them to a staff meeting and allow them the opportunity to conduct a training session on your alarm system. If you reach out and ask these contractors to assist you, you will begin to develop a partnership with these contractors. They will gladly provide training because an informed consumer is better equipped to understand the contractor and their advice.

As mutual respect is gained between consumer and contractor, a longer and more lucrative relationship is likely to develop. In today's ultra-competitive market, contractors are looking for ways to prove their worth, and you should take advantage of this.

Regular Testing

Even the most expensive and best designed alarm systems can be rendered useless if not properly maintained and tested. Time and time again, fire inspectors have discovered alarm systems that were adequate for the property at which they were installed, but because of improper maintenance, the alarm system did not operate properly and damage resulted. Businesses have been interrupted needlessly and in the worst scenarios, lives have been lost. The cost of maintenance is low, but the risk of not performing it is very high. The problem with maintaining an alarm system usually stems from two areas. First, the building owners or managers don't understand the need for maintenance and testing, or secondly, the incremental cost to maintain the system is cut from the operating budgets. Those responsible for emergency planning must ensure that neither of these two scenarios take place in your organization.

When evaluating your alarm system, it is important to know the age and condition of your system. A system that is over five years old should be examined for potential component breakdown. Normal wear and tear set in around this time, especially if the system is exposed to voltage fluctuations, humidity, dust, and temperature fluctuation. Periodic testing by a NICET certified contractor should identify these problems before they break down. NICET technicians are certified by the National Institute for the Certification of Engineering Technologies (NICET) and specialize in life safety issues. Technicians who hold NICET certification have a thorough knowledge of system installation and life cycle inspection, testing, and maintenance protocols.

If your system has not been inspected and maintained regularly, or if it has but only sporadically, beware. If the system is older than ten years, the equipment is susceptible to frequent and costly breakdown. Equipment installed over twenty years ago should be evaluated and possibly scheduled for replacement. The technologies used in your system may be antiquated and many of the parts may be unavailable. Because the purchase of a new alarm system can be costly, evaluate the current system and develop a five-year plan for replacement if the condition warrants.

> The National Fire Protection Association (NFPA) is an excellent resource for information about emergencies related to fires. The NFPA has developed a national fire alarm code (NFPA 72), and those responsible for emergency management should become familiar with that code. For detailed requirements of the code, visit the NFPA website at <www.nfpa.org>.

Most alarm system manufacturers recommend at least one full annual test and inspection; however, various agencies, organizations, and local authorities recommend—and in some cases mandate—more frequent testing intervals. Although not an all-inclusive list, the maintenance activities for fire alarm systems can be summed up in five basic steps:

- Test and calibrate
  Sensors such as smoke detectors must be tested and calibrated.
- Evaluate inputs and annunciators
  Place the system under test conditions and evaluate for problems.
- Set sensitivity
  Sensitivity must be adjusted based the system, location, and the potential emergencies. Specialized testing equipment is used for this task.
- Coordinate
  Coordinate with the fire department and the monitoring company to test inputs to their system and communication.
- Examine battery
  The battery must be examined for corrosion and other visual indicators. Check expiration date and test the battery.

The National Fire Protection Association (NFPA) provides the National Fire Alarm Code (NFPA 72). This standard covers the application, installation, performance, and maintenance of protective signaling systems and their components. Included in this code is information devoted to inspection, testing, and maintenance. The National Fire Alarm Code does not address whether a fire alarm is required in a given building; it simply addresses the alarm post-installation.

In addition to NFPA standards, OSHA has developed minimum requirements that often mirror the NFPA code. These OSHA regulations can be found in 29 CFR 1910.165. Furthermore, Appendix B to Subpart L of 1910 provides a cross-reference table that matches OSHA regulations to other applicable codes including NFPA and ANSI. An important point to remember is that in virtually all cases, these standards represent minimum requirements. These standards may be superceded by local jurisdictions and even manufactures' recommendations. Even if you meet the strictest of standards, you may still not have adequate protection for your facility. Unique issues such as voltage fluctuations, weather, dust, humidity, salt water exposure, and other environmental concerns can shorten the life of your system or otherwise affect the system in a negative way.

Hot Sites, Cold Sites, and Contingency Centers

Consider the aftermath of the September 11, 2001, terrorist attacks on the World Trade Center. From a real estate standpoint, nearly 30 million square feet of commercial space was destroyed or rendered unusable. Well over 1,000 businesses were affected, and four organizations lost over one million square feet of office space each, including Merril Lynch (3.1 million square feet) and American Express (1.2 million square feet). Where did these organizations go? Unfortunately, many companies closed their doors for good. Others found it necessary to find space in New Jersey or other surrounding states.

Imagine the mad rush for real estate. Imagine also the lost time as the office space had to be built to the specifications of the tenant. In many instances network cable, security, telecommunications, and office furniture had to be purchased and installed. Because of this dilemma, many organizations have turned to real estate companies that provide flexible office space. These real estate providers offer both cold and hot contingency sites. The cold sites include little or nothing in terms of technology and furniture. A cold site can be made ready with a short notice, but its not feasible to move into immediately. A hot site, however, can be moved into immediately. It is built to the tenant's specifications, including offices, cabling, computers, telecommunication, and furniture. In some cases the flexible office space is leased and used by temporary personnel or as swing space during renovations. If an emergency were to occur, the space could be quickly reconfigured and made available to core business units and functions. Other organizations simply lease the space and don't move in. It is used only if needed. Management realizes that the expense is necessary, and even if underutilized, it is there if needed. They view the space in the same light as an insurance policy. This may be a solution worth exploring if your organization's real estate is in one location that, if a regional emergency occurred, would cause an interruption in business continuity. In its most effective (and expensive) variation, a hot site might be built to provide redundancy. A redundant site is a secondary site configured exactly like the primary site. This includes the same furniture, computers, telephone systems, and other equipment.

Another possible solution is to create reciprocal agreements with other organizations. In other words, your organization would promise to provide a predetermined amount of square footage to a partner organization, and they would promise to do the same. This agreement may be the least costly, but it is often the most difficult to make work. Even if an agreement is reached, parties may fail to live up to their end of the deal because of any number of extenuating circumstances.

The Command Center

In the event of a major emergency, especially during the recovery stage, a command center must be established. This command center may be on site or off site if the emergency location is dangerous or if multiple locations are involved. Based on the emergency event and the location, the command center should be easily setup and transferred. For example, let's assume you manage 20 buildings in a three-county area. Prior to an approaching hurricane you decide to create a command center to serve as home base. It would be wise to have multiple potential command center locations so that if the hurricane hits your north county property, the command center can be moved to the south county property. This would serve to keep the command center from being damaged and would preserve utility of the center.

Off-site Storage

Now more than ever before, data security is paramount. Given the role it plays in business continuity, companies can ill afford to be without it. In many cases, off-site storage of data is a necessity. To understand this, think about a disaster event like the World Trade Center attacks. Because the buildings were leveled, any data stored on site were lost.

When considering backup options, solutions such as magnetic tape, CD-ROM, microfiche, and digital formats exist. These decisions should not be made without fully understanding the strengths and weaknesses of each. Those issues cannot be adequately covered in this book, so consult with information technology professionals who can evaluate your needs and your unique situation. One final word about electronic data: consider the possibility of data corruption due to virus infection, sabotage, and other problems.

Data Security

Data security is a science unto itself, and certainly the pages of this book are unable to do the topic justice. This is not a book on disaster planning in a digital environment; however, a few words are in order.

- How often is the server backed up?
- Are employees saving documents to the central server or local hard drive?
- How long is the back-up data kept?
- Is off-site storage used?
- What physical security precautions are implemented at the server rooms?
- What would happen if the server rooms were damaged?
- Is a fully operational hot site necessary?
- Are we protected from viruses?
- Are we protected from hackers?
- If data were lost, how quickly can it be restored?

In a world so driven by digital technology, loss of computer functions can be devastating. Contingency strategies must be developed and implemented to protect the digital integrity of your business. Typically, businesses protect themselves by performing regular data backup and securing off-site storage, but is this sufficient? In the wake of the 2001 terrorist attacks, we've had to rethink our contingency efforts. We've always planned for every possible contingency including terrorism and airplane crash, but we've never planned for the scope that occurred on 9/11.

After the World Trade Center collapse, the New York financial district was in shambles. Thanks to aggressive pre-planning and sheer determination, the New York Stock Exchange was back online in just four days. The NYSE had spent over $20 million on a contingency site complete with an alternate trading floor in the event the New York location was damaged. Trading firms also had contingency sites in place, some of them outside of the state. One of the greatest concerns for any contingency site is the need for seamless communications. What systems are in place to ensure telecommunications will be available? Will computers operate as planned? Can backed up data be loaded on the hot site computers in a quick fashion? How quickly? If business continuity is critical to your organization, consult with your information technology employees and discuss available contingencies. It will likely be necessary to bring in specialized assistance that can match your needs with the proper contingency strategy.

Lack of Electricity Doesn't Mean "Go Home"

Loss of utility is an ever-increasing concern. Although rarely considered a disaster (or even an emergency), loss of utility can negatively impact an organization's productivity. Even momentary blips in electrical service can result in loss of digital data and disconnection of telecommunications. A number of solutions exist and can be realistically implemented to solve the problem of loss of electricity. Like any solution however, pros and cons exist and must be addressed prior to making the investment. A good starting point is to identify the areas of your business that can least afford to be without electricity. These functions can be identified as "mission critical" functions.

Identifying Mission Critical Functions

Suppose your organization produces widgets. Your factory has an assembly line that produces 100 widgets per hour. Often, your line workers find themselves up against critical deadlines. For each hour that the assembly line is unable to produce widgets, your organization losses $10,000. It's very easy to identify the production floor as a mission critical function. Other times, it is not so easy to identify the mission critical functions of an organization. The problem usually occurs when every business unit believes it is equally important. When an organization cannot prioritize the critical nature of its business, problems will exist. For this reason the emergency planning effort needs to include members of the executive management function. They will be able to provide direction and clarification concerning the strategic direction of the organization. Too often, if the task is left to mid-level managers, critical functions become synonymous with whatever those managers are in charge of. We tend to place more importance on our jobs, perhaps not understanding what other departments do and how that affects the organization's business plan. In the example of the widget producing factory, peripheral tasks such as human resources and marketing would not be considered mission critical tasks. This is not to say these tasks are not important. We are saying you should prioritize which business functions need to be restored first; human resources and marketing would come after production.

When evaluating business continuity and mission critical functions, some thought should be given to options that are available in the event of loss of utility. For example, based on your organization's business, is it too expensive to purchase and maintain an emergency generator? Is the probability of loss of utility a small one? In the event of prolonged loss of utility, would it make better sense to move processing to an alternate site? Are the consequences of a loss of information technology resources sufficiently high to justify the cost of various recovery options? Performing a risk assessment will allow you to focus on areas in which it is not clear which strategy is the best.

Uninterruptible Power Supply

When loss of electrical utility is experienced, the event usually lasts for a very short duration, just minutes or often seconds. However, the impact of the momentary loss of electricity can sting for a long time. If a computer server goes down, even for a minute, it will affect everybody routed through that server. Because of the lost productivity and the associated costs, information technology professionals view uninterruptible power supplies (UPS) as a necessity. The UPS system could be an inexpensive localized unit that supports the single computer plugged into it, or the UPS system could be a highly advanced and expensive unit that supports a full room of computers and equipment.

When considering a UPS system you must first understand what they can do and what they cannot. UPS systems perform two main functions. They supply short-term power to equipment during a loss of utility, and they compensate for electrical surges and sags that could damage hardware and corrupt data. Most UPS systems provide less than fifteen minutes of power, mostly via a battery power source. If a power outage extends beyond the backup capacity of your UPS system, you will have had time to save your work and properly shut down the computer equipment.

If you have the need and the budget, you can purchase UPS systems that provide power to a room full of computer equipment for an hour or longer. Its important to realize that UPS systems are not the best solution for powering up an entire building during an extended power outage. If you have a need for high-load, long-term power provision, emergency power generators should be considered.

Emergency Generators

An emergency power generator has the ability to provide electrical power over an extended period of time—hours and even days. When power into a building is lost and a generator senses the loss, it will begin to start up and transfer the power source from the utility feed to the generator. Because of the time involved in this process, the emergency power generator is not good at regulating power quality or bridging momentary power blips. A good emergency power plan usually includes both a UPS system for short-term power loss and surges and a generator system that picks up where the UPS system leaves off.

The first step when considering the purchase of an emergency generator involves the sizing of the unit. Building occupants often mistakenly believe that the generator will supply power to the entire building. They falsely believe that all building systems including heating, cooling, and lighting will be operational. That is just not the case. To purchase a generator that would power an entire building during those rare moments of loss of utility would not likely be cost effective. Because of this, a risk analysis should be conducted prior to specifying and purchasing an emergency generator.

Begin by walking the building and taking into account the type of equipment present. Speak to the building occupants to gain a thorough understanding of their business. Work with these employees to whittle down the list of mission critical equipment. Each employee will want their desktop printers and fax machines to be operational during a power interruption. They will insist upon properly working heating and cooling equipment. They will absolutely demand that the copy machine in the library be operational. And don't forget about the refrigerator and microwave in the break room. You will need to use your power of persuasion and reasoning, along with a good dose of political tap dancing, to help these employees understand the purpose and limitations of the emergency power generator. You can approach the generator load factor in two ways. The first and perhaps most realistic option is to identify a budget for your generator purchase. Based on this budget, you will be able to afford an "X" kW generator size. Let's assume your budget will allow the purchase of a 100 kW generator. You now have a limited bucket of power available to your building occupants. You would work with them, getting their suggestions and support when deciding what equipment will be placed on the generator. The second option is to first work with your employees in deciding which equipment must be on the generator. Once this list has been compiled, you would then identify the size generator needed and budget accordingly. As you might imagine, option two will always be the more expensive option.

When evaluating generator options, you will be faced with several engine and fuel options. The most common options are diesel, gasoline, and natural gas. Each of these options has advantages and disadvantages, and you must take these into consideration. Most of the differences revolve around cost, fuel, and size.

Diesel

Positives:
- Excellent reliability
- Typically available in larger sizes than gasoline or natural gas
- Diesel fuel has a long tank life (if tested and treated properly)

Negatives:
- Can be more expensive than gasoline or natural gas
- May be difficult to start in cold weather
- Cold weather units require oil heaters and engine block warmers
- Diesel fuel must be tested and treated for water absorption

Gasoline

Positives:
- Lowest initial price
- Easy to start and run, particularly in cold environments
- Available in a range of sizes up to 100 kW

Negatives:
- Gasoline has a relatively short tank life; gas must be treated

Natural Gas

Positives:
- These generators are less expensive than diesel (slightly more than gasoline)
- Fuel has a long tank life
- Available in a range of sizes up to 100 kW

Negatives:
- Cost and availability of fuel could be an issue

The installation of an emergency power generator is no small project. In addition to the expense, you must have adequate space to place the unit. In addition to the generator, space will be required if you plan to install an automatic transfer switch. Because of the size, noise, and vibration that is inherit with an emergency power generator, an exterior installation usually makes the most sense. One of the common mistakes made when locating a generator is placing the generator and its exhaust near fresh air intakes or windows and doors. The exhaust generated can seep into your building and create a serious indoor air quality problem. You won't realize you have a problem until the generator is running.

> There is a big difference between price and cost. Price describes the initial investment required to acquire something. Cost is used to describe the initial investment plus the cost of maintaining the equipment over its useful life. Often the lower priced option is the higher cost option over a period of years. For this reason, understand the value of the entire investment, not just the initial purchase price.

In some scenarios local utility providers will advise its customers to limit electricity usage. This is called "load shedding." In a practical sense, load shedding occurs when people raise their thermostats and turn off unnecessary items such as escalators and some lights. The load shedding process is usually a precursor to emergency power generator usage. If you have an emergency power generator, you should also develop emergency power procedures. These procedures will become part of your emergency plan. In addition to load shedding procedures you may include these issues in your emergency plan:

- If your generator does not include an automatic transfer switch, who will be responsible for switching generator power on?
- Who will confirm that generator power has transferred properly?
- Who will the building occupants call in the event power does not transfer properly or the generator does not start?
- Who will provide regularly scheduled maintenance on the generator?
- What will the cost of regularly scheduled maintenance be?
- Who will be responsible for checking and ordering fuel for the generator?
- Has a purchase order and contract been initiated for fuel delivery?
- What additives must be used for the preservation of fuel?

After the initial investment in an emergency power generator, you must spend time and money maintaining the generator. Much like an automobile that requires an oil change every 3,000 miles, your generator requires regularly scheduled preventative maintenance. In fact, a generator engine is very similar to your automobile's engine. The generator has a battery, cooling system, belts, hoses, oil, plugs, and so on. Each of these items has a useful lifespan and must be maintained. Given these facts, you must initiate a preventative maintenance plan. Since most generators are installed outside and exposed to the elements, make sure that vegetation is trimmed back from the generator and leaves and other debris are not blocking the airflow to the generator's radiator.

A less expensive generator option might be to have a portable generator delivered to your location. These portable generators are not the micro-sized units available at the hardware store. These units can be as large as any permanent generator, except they are mounted on a flatbed truck or semi and delivered to your location when needed. Perhaps after performing a risk analysis you decide the price of the generator is not worth the investment, considering the occasional loss of utility. You would, however, like to protect your business continuity in the event of a major, long-term loss of utility such as

> Having a generator is great! Having a generator that actually starts up when you need it is essential. Unfortunately, many generators don't start when we need them the most. What is the most common cause of generator failure? Believe it or not, most generators don't start because of a dead battery! The second most common reason generators don't start—lack of fuel or poor fuel quality! These problems can be easily solved if you have a regularly scheduled maintenance plan in place. It is essential!

a that brought about by a storm or earthquake. If you elect to use this option you should consult with a generator vendor and ensure that a portable trailer-mounted generator will be available. This will often necessitate you purchasing an option on the generator. Don't take the vendor's word that they will have a generator for you. You must take steps to ensure the vendor will guarantee delivery. The contract must clarify issues such as

- What size generator is needed?
- Who will wire the generator to the building?
- Is this building able to accept the generator?
- Who will be responsible for initial fueling and refueling?

This agreement will require the generator provider to visit your building and understand your needs. Just as if you were purchasing a generator, you must properly size it up. You must determine which circuits will be powered by the generator. This may require an electrician to rewire or add electrical circuits. Typically, receptacles that are plugged into generators are identified with a unique color, usually orange. These issues must be addressed prior to the delivery of the generator. If an event warrants the generator delivery, it will be too late to resolve these critical issues.

As a last option, depending on your recovery and continuity needs, the micro-sized portable generators may be an option. These units can provide small amounts of electricity in an emergency, but don't count on these units to power up your building. They may be sufficient to provide limited lighting at night or cool a refrigerator to prevent spoilage. Understand their limitations and prioritize your needs. As a safety reminder, never plug the generator into the wiring at your circuit breaker. This can cause a backward flow of electricity, which could endanger utility personnel working on the power lines. Also, keep the generator in a well ventilated area outside of the building. Exhaust fumes can be deadly, especially in gasoline-fueled generators.

Summary

The Occupational Safety and Health Act requires that businesses address certain types of emergency events in their organization's safety plans. Specific elements include fire plans and evacuation procedures. Additionally, OSHA requires you to evaluate your business and address the hazards that are present. These potential hazards may include flooding, chemical release, hurricanes, or earthquakes. Failure to plan for these events may result in an OSHA citation and fine.

When developing your emergency action plan, the identification of key contractors and employees is critical. These key contacts can help you both during the emergency event and during the restoration phase. The creation of a chain of command is also critical. Who will make the decision to evacuate or remain in place? Who will be asked to stay behind and shut down the building? Who will serve as floor captains? The time to answer these questions is before the emergency event occurs.

6

Practice Makes Perfect: Training and Drills

So You Have a Plan. Now What?

Congratulations! You've finally created the perfect emergency action plan. A lot of hard work went into the design of this plan, but a good plan is only half of the equation. Distributing the plan and fostering employee support is also required. Distributing the plan is not too difficult, but how do you get employees excited about this new process? That will be discussed a little bit later, but first let's review the eight steps in plan development:

- First draft
- Review
- Second draft
- Tabletop exercises
- Final draft
- Print and distribute
- Employee training
- Annual review and revisions

The first three steps are often the most difficult. When developing the plan, understand that the first draft is not an exercise in perfection. It's a starting point, a launching point, if you will. The second step is the review process. This step can be discouraging because you'll be poking holes in step one. When conducting the review, ask participants to "bring me solutions, not problems." This encourages constructive ideas instead of just random complaining. It's okay to play devil's advocate. Ask tough questions. If you don't understand a process, say so.

The second draft will be a better representation of the finished plan. It's not perfect, but it's getting close. The second draft is tested for validity through tabletop exercises. This is often known as the fire testing stage. Scenarios are drawn from a hat and the emergency management team walks through them using the plan as a guide. Any conflicts or gaps would be corrected in the final draft.

At this point the plan has been battle tested and is ready to go. Printing and distributing the plan is a precursor to the most critical stage in this eight-step model—employee training. Employee training may consist of such diverse activities as electing floor captains, performing evacuation drills, or attaining CPR certification. A great emergency plan is worthless without employee involvement. Conversely, a good emergency plan supported by employee participation is worth its weight in gold.

Of course, writing the plan is an ongoing process. The emergency plan should be considered a living document, subject to growing pains and multiple revisions. A minimum annual review of the plan is a necessity.

Rolling Out the Emergency Action Plan

Once the plan is finalized and all revisions have been made, the plan should be distributed to those playing a support role. These people might include emergency team members, floor captains, fire brigade members, evacuation assistants, runners, command post leaders, and anyone else who may be asked to play a key role in the emergency recovery efforts of the company. A good time to distribute the plan is at a training workshop for those employees who fill support roles.

To add credibility to the initial training workshops, open the meeting with a word from the company president or some other high-ranking management member. This will impress upon those in attendance the importance of the roles they play. Lives are at stake. Business continuity is at stake. The organization's reputation and productivity can be affected in a good or bad way. If this is true, back it up by a quick word of thanks and encouragement from management. It really does set a positive tone for the planning and recovery team.

The second reason for the training meeting is the opportunity for key support members to meet one another. Since the group will be a diverse one, many of the members may not know each other. Your goal should be to cultivate an atmosphere of trust and reliance. This only comes when the members of the support team know and believe in one another. One of the best ways to cultivate this atmosphere is to conduct team exercises. There is no need to resort to corny team building exercises, just have the members of the support group work on "what if" scenarios with members of their team. Of course, growth will not be experienced if everybody pairs off with people they already know, so suggest that they join a group of employees they don't know very well.

Finally, make the training enjoyable. By providing give-aways such as movie tickets and lunch certificates you'll gently nudge participation. To highlight the importance of this training and the supporting employees, consider awarding everyone present with a special golf shirt or coffee mug identifying the bearer as a member of the team. Each of these ideas will help foster credibility and stress the importance of the roles being filled by the members.

Training the Emergency Action Team

Because the emergency action team will be taking the lead in the event of an emergency, team members must be familiar with the plan and the subject property. Both considerations will take time and training. The training will not be a one-time event. It will be an ongoing event with a goal of preparing those taking the lead. Specific training formats include the following:

- Orientation training should be delivered to new members of the emergency action team. Each member should understand not only their responsibilities but also the responsibilities of the other team members. At the orientation, team members should be assigned a copy of the emergency action plan book. The book will be their personal property, and if they leave the team or the company, they must return the emergency action plan. The team member will be responsible for updating the plan as revisions are distributed.

- Tabletop exercises are an excellent way to train members of the emergency action team without disrupting the normal day-to-day activities of the organization. In a typical tabletop training exercise, members of the team sit in a conference room and pull emergency scenarios from a hat. Circling around the conference room table, each member would then explain their responsibilities to the group. A second pass around the table would allow team members to ask questions of other team members. If a weakness is discovered or an opportunity to improve is noticed, the entire team would work together to improve the emergency plan.

- Walk-through drills allow the emergency action team to more accurately perform their emergency responsibilities. During a walk-through drill, members of the emergency action team visit areas that are crucial to a successful plan implementation. Such areas might include shut-off valves, alarm panels, automated external defibrillators, first aid stations, and exterior employee staging areas. The walk-through drills are essential for building familiarity and can be performed without impacting normal business productivity.

- Pre-evacuation drills are an extension of the walk-through drills. Involvement in this drill is limited to members of the emergency action team. Participants are instructed to evacuate the building, paying particular attention to the signage, exit routes, doorways, and staging areas. This is a perfect opportunity to evaluate the evacuation process without impacting the entire workforce. Participants should quickly evacuate as if an emergency is present. During the evacuation, notes should be taken, but that may force the participants to stop and write down their observations. A solution to this problem would be to provide participants with small tape recorders. As they evacuate the building they would simply record their observations into the tape recorder without even stopping. After the exercise is completed, the participants would play back their recorded observations and make adjustments as necessary.

- Full-scale exercises are useful in many organizations. The purpose of a full-scale exercise is to simulate an actual emergency event as closely as possible. This necessitates the involvement of emergency response personnel, employees, and community response organizations. Because of the involved nature of the full-scale exercise, this option may not be easily implemented. You'll need the cooperation of a number of different agencies, and they are not usually able to fulfill all requests. As an option, you may wish to have your emergency management team participate in another organization's full-scale exercise. This option won't address your organization's specific needs, but it will allow you to "go live" and experience a staged emergency event. The experience gained can be of tremendous value and should not be overlooked.

The orientation training should be delivered to new team members at the time of their assignment. Thereafter, training should be provided to all team members at least annually. Other occasions when training, or at least an update, would be required include

- When a process or responsibility is revised
- When policies affecting emergency planning and recovery change
- When a change to the layout of the physical building occurs
- When new equipment is introduced that may affect emergency planning (i.e., a generator or chemical storage)
- When key personnel changes
- After each emergency

> Revisions to the emergency action plan will occur. It is critical that you track these changes and ensure that all owners of the plan have the correct information. You can accomplish this by inserting the page number, document title, and the revision date at the bottom of the page. This will allow you to quickly look at the emergency plan and determine if the information is up to date. As information is revised, simply print the newly revised pages and distribute them to your team. Be sure to leave three or four pages intentionally blank at the end of each section, so you don't have to reprint the entire manual.

Training the Employees

Training is particularly important for effective employee response during emergencies. There will be no time to check a manual to determine correct procedures if there is a fire. Depending on the nature of the emergency, there may or may not be time to protect equipment and other assets. Practice is necessary in order to react correctly, especially when human safety is involved. Although most employees will not receive the depth of training that the emergency action team members receive, they still need some training including new employee orientation, tabletop exercises, announced evacuations, and unannounced evacuations.

Emergency orientation training should be delivered to new employees as they accept assignments with your organization. This may include contractors such as janitorial personnel who spend substantial time on your property. Realistically, when a new employee starts work at your organization, they spend the first day or two filling out paperwork, reading through the company policies, asking questions about health insurance, and other related activities. The emergency action policies that you provide them will be their first exposure to the safety culture your organization deems important. You have the opportunity to help them realize the importance placed on employee safety and business continuity. Don't miss this opportunity! It is likely that you will need to work with your Human Resources department to ensure that your emergency policies become part of the standard new employee orientation process.

> When training adult learners, three factors are required:
>
> Awareness
>
> Education
>
> Involvement
>
> The program starts with awareness. Next, formal education must take place. Finally, if any of this information is going to be retained, you must get involvement. It's really that simple!

In most organizations, especially mid-sized to large companies, a complete copy of the emergency action plan will not be given to employees. The plan contains information that is not relevant to most employees, and if you overburden these employees with information, they will not look at it. Instead, provide employees with a scaled-down version of the plan, perhaps an executive summary with important phone numbers and a simplified floor plan.

Tabletop exercises are an excellent way to train the general employee population without disrupting the normal day-to-day activities of the organization. In these tabletop exercises, many excellent ideas and improvements have come from the employees. Typically the employees will challenge the members of the emergency action team with questions and issues that were not discussed previously.

Announced evacuations are a great way for the employees and the emergency management team to gain confidence in one another. The problem with most evacuation drills is the timing. People are busy and the evacuation is rarely considered a priority. When the evacuation drill does occur, people are confused, meetings are interrupted, deadlines may be impacted, and more time is spent performing the evacuation because of the unfamiliarity with the procedures. By announcing the evacuation, you allow the employees to prepare for the event. You also allow the employees to find the emergency exits prior to being evacuated. You have a chance to explain the evacuation routes, staging areas, and other related procedures. Eventually, as the employees become more familiar with the evacuation procedures, they will be ready for the unannounced evacuation.

Unannounced evacuations more accurately simulate an actual emergency event. When faced with an emergency we often forget the obvious. Where do we go? Where is the closest emergency exit? Where is the secondary exit? Who needs assistance in evacuation? Where is our staging area? Who performs the employee's roll call? Questions like these are easy to answer during a tabletop exercise, but when the unannounced evacuation occurs, we often get rattled. By conducting unannounced evacuations you provide employees with the opportunity to work out the kinks. The unannounced evacuation should not be conducted until your employees are comfortable with the announced evacuations.

Fostering Employee Support and Enthusiasm

Some organizational programs are easy to sell to the employees. If a new and improved benefits package is being unveiled, employees are eager to listen. They can see the tangible benefits of this new package. As the package is being explained, employees will be on the edge of their seats, paying rapt attention, even asking thoughtful questions. Now take a less popular subject and see what kind of response you'll receive. Traditionally, emergency planning and recovery has been one of those less-than-glamorous topics that send people scurrying for the doors. This is a real problem because a great plan without employee support has little value.

Given today's realities, fostering employee support (and even enthusiasm) for emergency planning and recovery should be easier than in years past. It will still require effort. Excitement must be generated and nurtured; but how do you do that? When unveiling the emergency plan, start by letting the employees know that one exists! That's a logical starting point, but how do you make it happen? Management, at its highest levels, should take the lead in trumpeting the emergency plan. This announcement can be delivered via e-mail, posters, newsletters, or any other medium that employees are likely to listen to. The e-mail example is a good one for a few reasons. It's inexpensive to send. One e-mail message can be sent to thousands of employees. Think of the impact of receiving an e-mail "from the desk of R. A. Smith, President, Acme Industries." Would you read it? I know I would!

The next logical step is to provide employees with training classes. Until you educate them about the plan, how can you expect them to get excited? Don't overwhelm them. Don't attempt to teach them everything you know about emergency planning and recovery in two hours or less. Provide the training in small increments. Keep the information specific to what they need to know, taking into account that not everybody needs to understand all of the details. For those volunteering for additional responsibilities such as fire brigade or floor captain, you'll provide them with additional training pursuant to their responsibilities.

Keep the training light, informative, and fun. You don't have to play silly games to get their attention, just keep them involved. Role playing emergency scenarios is a good way to keep employees interested. People like to get rewards, so why not bribe them? Inexpensive trinkets such as golf balls, lunch certificates, or movie passes can be awarded to participants who answer questions correctly. With a little preparation you'll be able to roll out the emergency plan with excitement and anticipation.

Getting Everybody Out Safely

Make no mistake about it; evacuations are performed to preserve the organization's greatest assets: its people. We don't evacuate computers or furniture, only people. Wouldn't it be tragic if, in an attempt to keep our employees from harm, we evacuated them from the building, but they were injured during the process? To protect our employees from injury, we need to establish a logical means of egress from the building. Evacuations are not an easy task. Sure, we exit the building each day, but we do so without duress. We follow the same predictable routines. The elevator quickly moves us to the lobby. We exit and make our way to our automobile. An evacuation forces the building occupants to do something entirely different. Because the elevators will be off limits, where will they find the emergency exits? What about employees with physical limitations? Is the exit route so intuitive that visitors will be able to exit? Where do they go once they are outside? These are the types of questions we will begin to answer.

Evacuation Routes and Drills

Regular evacuation drills are a critical component of a successful emergency action plan, but who can afford the impact of evacuating an entire building during a single drill? Here's the good news: a total evacuation is not necessary, nor is it recommended. Smaller drills have proven to be quite effective, and they provide the training necessary for the employees to gain confidence in the event of an actual emergency.

A full evacuation of high-rise building is difficult thanks to a number of different factors. First, it's very time consuming. A large building typically contains thousands of people and it may take hours to get them out safely. This is especially important to note because most fires are localized to a floor or two. The NFPA has embraced an evacuation model that conducts a partial evacuation of the affected floors and the floors immediately above and below the affected floors. For most fires in most buildings, this course of action is sufficient. Of course, some extreme emergency events such as the World Trade Center bombing warrant a full evacuation, but that is the exception to the rule.

Consider also the problems of re-entering a large building after a total evacuation. It typically takes even longer to re-enter a building. It's one thing to instruct employees to walk down 70 flight of stairs and quite another to ask employees to walk up 70 flights of stairs. Because of this, most employees will wait for the elevators to take them to their assigned floors. Buildings were not designed to accept an entire occupancy in a short period of time. Organizations within a large building usually stagger the arrival and departure times of their employees so as not to overburden the building's elevators and common areas. A mass re-entry into the building works against this plan and creates a large problem.

> When developing evacuation plans for any building, especially high-rise buildings, you must take time to consider the results of any decisions made by the emergency management team. One of the best ways make this happen is to follow this simple advice:
>
> Before you conduct an evacuation drill, do a walk-through. Before you have a walk-through, have a talk-through.

A final reason to avoid the entire building evacuation approach is the negative impact on business continuity. This is a secondary consideration, but a consideration nonetheless. When a small fire directly affects two or three floors, does it make sense to evacuate 70 floors? Doing the latter would affect all of the building's occupants, resulting in a tremendous loss in worker productivity.

A common misconception when dealing with fires in a high-rise building is that it is safe to travel up toward the roof and await a dramatic helicopter rescue. True, this feat has been accomplished, but it is very dangerous and not recommended. Performing this type of rescue puts the helicopter pilot, the building occupant, and the fire fighters at risk. If the fire is severe enough, large thermal currents will rise from the burning building making the helicopter hard to control and the downward thrust from the helicopter's rotors can force smoke and hot air back down onto the fire rescue personnel who may be close by. Because of the danger involved, many pilots will not even attempt this sort of rescue. Individuals who wait at the rooftop are often left behind.

In light of perceived vulnerability of high-rise buildings in an emergency event, many local municipalities have developed additional safety requirements. The city of Chicago has recently enacted such rules and is recognized as being at the forefront of high-rise safety.

> Always be aware of local agencies that promulgate evacuation protocol and codes. The best way to learn about these issues and expectations is to invite your local fire marshal to visit your building and offer suggestions.

Some of the requirements of Chicago's emergency procedures (Title 13 of the Municipal Code of Chicago: Chapter 13-78) for high-rise buildings include issuance of an "emergency preparedness certificate" by the fire department upon receipt of adequate proof that the applicant is able and qualified to assume the duties required.

High-rise buildings are grouped into four categories based on height above grade. For instance, Category 1 structures are over 780 feet while Category 4 structures are 80 feet and over, up to and including 275 feet. Some of the categories of buildings are required to file a copy of the safety plan with the city's office of emergency communications.

Each emergency plan for Category 1 buildings must include the following required designated personnel:

- The Plan must designate a Fire Safety Director (FSD). The FSD must be an employee of that building. The FSD shall obtain and maintain an emergency preparedness certificate. A deputy Fire Safety Director may also be required.
- The Plan must designate a Building Evacuation Supervisor.
- The Plan must designate an Emergency Evacuation Team. Emergency Evacuation Team members shall know the location of all exits and lead emergency evacuations and drills as directed by a Fire Warden.
- The Plan must designate Fire Wardens in sufficient number to carry out their duties. The wardens must know the locations of all exits and direct emergency evacuations and drills from their assigned floor in accordance with the Plan.

In the event of an actual fire or emergency, the Fire Safety Director and Deputy Fire Safety Director must occupy the building's fire command station, conduct emergency operations, direct evacuations, and report conditions to first-arriving fire companies. Regular duties include

- Conduct monthly building safety inspections
- Develop procedures for emergency evacuations and drills
- Direct emergency evacuations and drills
- Assign Fire Wardens and assign Emergency Evacuation Team(s), if required

As part of the minimum requirements of the plan, the actions all occupants should take in an emergency evacuation or drill should be described. In accordance with commonly accepted fire evacuation methodology, each plan shall set out a procedure for an evacuation of five floors below and two floors above any emergency resulting from a fire on a certain floor. Additionally, the plan must specify procedure for a full evacuation of the building. On certain occupancy classifications of Category 1 high-rise buildings, safety drills must be carried out under the direction of the FSD not less frequently than twice a year. Furthermore, on an annual basis the owner shall file an affidavit with the Fire Commissioner certifying that at least two safety drills have taken place on all occupied floors during the past year.

Each plan must list the name and location of each occupant who has voluntarily self-identified that they need assistance and the type of assistance required to swiftly exit the high-rise building in case of an emergency. This emergency plan must be kept in the office of the high-rise building, at the security desk, and in the vicinity of the fireman's elevator recall key or life safety panel. The development of specific emergency planning regulations for high-rise buildings is quickly becoming the norm throughout the country. The forced awareness of emergency events and how they should be handled can only serve to save lives in the future.

Organizations may also arrange for on-site disaster simulations. These simulations are acted out with the utmost detail, including fake blood and wounds. Local emergency response teams are called in to participate. These simulations provide valuable information about flaws in the contingency plan and provide practice for a real emergency. While they can be expensive and time consuming, these tests can also provide critical information and expose weaknesses in a plan. The simulations are not suggested for every organization, but the more critical the functions of your business and the resources addressed in the contingency plan, the more cost beneficial it is to perform a disaster simulation. These simulations are often performed at healthcare organizations and state or local government agencies.

Minimizing Interruptions and Down Time

One of the problems with building evacuation drills is the loss of productivity. Businesses have deadlines to meet and an evacuation drill can impact those deadlines. If the evacuation, staging, roll call, and re-entry take an hour and one hundred employees are evacuated, the organization has lost one hundred hours of productivity. A further case against evacuation drills can be made if you attach a dollar amount to that lost productivity. For instance, assume the average billable dollar amount per employee is $25. Multiply that billing rate by the 100 employees and the cost to conduct that drill is $2,500. These are realistic concerns, but they should not prohibit you from conducting evacuation drills. The first question to ask is "What is the value of a human life?" Is it $2,500? Is it $25,000? Is it greater than that? It's impossible to realistically place a value on a human life, but I think we can agree $2,500 does not come close. The emergency manager is not oblivious to the loss of productivity. The emergency manager will take steps to impact the employees as little as possible. Providing management with an idea of when the drills will take place gives them an opportunity inform you of conflicts. If a major corporate deadline is looming and employees are working overtime, perhaps the emergency evacuation drill can be deferred until next quarter. By training employees prior to the evacuation drill, you'll not only conduct a more efficient drill, you'll also finish it more quickly, thus impacting the employees' productivity less.

A full-scale evacuation of a building, especially a large building, is often difficult and time consuming. Many organizations have successfully devised a method to conduct emergency evacuations on a much smaller scale. Instead of setting the fire alarm throughout the building, an emergency manager may elect to enter a specific department and, using a whistle or megaphone, announce the fire drill. Only the building occupants in that work area would respond to the alarm. If the building occupants are likely to head for the obvious and closest fire exit, try adding a new component. Place an easel at the closest fire exit with the words "Do Not Enter—Fire Burning Here!" written on flip chart paper. Observe how the building occupants respond when they see the sign. Do they know where the next emergency exit is located? Regardless of how the drill is handled, this will provide the emergency planner with an excellent evacuation topic to address.

Healthcare facilities present unique but not impossible evacuation challenges. It's not likely the emergency planner will walk to a fire alarm pull station, trigger the alarm, and expect to evacuate the building occupants. Given the sensitive nature of the work being conducted in a healthcare environment, it is impossible to conduct evacuation drills. How can healthcare employees be trained in proper evacuation techniques in the event of an actual emergency? In the healthcare industry, tabletop exercises have largely taken the place of full-scale evacuations. Employees gather in a conference room and are given emergency scenarios. The employees must discuss the sequence of events following the emergency. Which sections of the building should be evacuated? What alternate sources of power are available? How will employees communicate if phone and electric service is lost? Where can we safely move patients in the event of a fire in the east wing? These types of questions will provide mission critical employees the training they need without negatively impacting the safety or quality of their work. These same tabletop exercises can be used in a number of different scenarios, including tall buildings in which it would be impractical to evacuate employees from the 76th floor.

Alerting the Building Occupants

OSHA requires employers to devise a method for notifying employees of an emergency.

In most scenarios this is accomplished via an audible alarm that can be triggered manually or automatically. But think for a moment about the problems with an audible alarm. Those with hearing disabilities may not hear them. In a loud shop area, the alarm may not be heard over the noise of the equipment. When evaluating your alarm system, make sure you take these issues into consideration. If excessive noise or persons with hearing limitations are present, you should devise a secondary alarm warning system. For instance, an audible alarm coupled with a visual alarm such as a

> Many hotels offer their visitors the option of a room with an audible alarm and a visual strobe. Hotels with a comprehensive evacuation plan have realized the importance of keeping a list of visitors with hearing impairments and, in the event of an emergency, would take special steps to ensure they are aware of the alarm.
>
> In some government buildings, persons with hearing impairments are given a pager that vibrates when an alarm has been activated.

flashing strobe would provide much better protection. Also consider areas in your building that may not be covered by the audible alarm. These areas might include basements, restrooms, and non-renovated portions of the building.

OSHA also suggests that distinctive alarm tones may be helpful. For instance, if your building has an organized fire brigade, they will need to know if the alarm is for a fire or if it is for another emergency that does not require them to assist. For a fire, the alarm may be a siren while an alarm for an approaching tornado might be a series of horn blasts. Its not recommended that you have a distinct alarm tone for each type of emergency. Given the number of different types of emergencies, it would become difficult to remember what each alarm means.

In most cases, the building occupants will discover the emergency. For this reason, the employer must explain to the employees the preferred means of reporting emergencies. Methods may include manual pull box, public address systems, radio, or telephone. If any method beyond a manual pull box is required, the employer should train the employees in proper procedures. Who should they call? What radio frequency should be used? How does the public address system work? What areas does it cover?

If a public address system is used to notify the building occupants of an emergency, you should have a pre-written script ready to read. This script would explain the actions necessary. It should be kept short and without detail. Don't work the group into a frenzy. The language that you use can be calming and helpful, or it can be confusing and frightening. The latter (not recommended) is usually the result of not having a prepared script. A sample script might look like this:

(READ, REPEAT, WAIT 90 SECONDS, AND REPEAT AGAIN)

"We have received an alarm in _____ building. As a precaution, we are requiring evacuation of all employees in _____. Please immediately proceed in an orderly fashion to your assigned exterior staging areas."

In this hypothetical situation, it has been decided that a partial evacuation is all that is necessary, so we would fill in the blanks with the pertinent information and read it over the public address system:

"We have received an alarm in <u>the north wing of the</u> building. As a precaution, we are requiring evacuation of all employees in <u>the north and west wings</u>. Please immediately proceed in an orderly fashion to your assigned exterior staging areas."

Notice that this script does not provide space to discuss the details of the alarm. Does it matter if the emergency is a fire, bomb threat, or chemical spill? They are all emergencies, and the evacuation should occur immediately and in an orderly fashion. By providing details, you risk frightening the building occupants. When this happens, people get careless, run for the exits, and those with disabilities are left behind. If you've conducted evacuation drills in the past, the occupants will know exactly what to do; there is no need to scare them into leaving the building. Notice also the language used. The second sentence begins, "As a precaution...." This is calming language meant to reassure the building occupants and remind them of their safety. However, at the risk of this alarm sounding unimportant and perhaps voluntary, we follow up with "we are requiring evacuation of...." The last sentence encourages the building occupants to "immediately" and "in an orderly fashion" leave the building. Finally, a reminder is given to go to "your assigned exterior staging areas." This is a good all-purpose script that can be adapted as you see necessary. A variation of the above script is to change the last sentence to read:

> **"Please take your personal belongings and immediately proceed in an orderly fashion to your assigned exterior staging areas."**

This change is recommended in bomb threat scenarios. By reminding the building occupants to take personal belongings such as purses, briefcases, duffel bags, jackets, and lunch boxes, you remove many of the items that would have to be searched during the bomb sweep. This facilitates the bomb search and solves the problem of possible liability when searching personal items.

The verbal instructions given should be understood by all building occupants. This becomes a challenge when some building occupants don't understand English. In some areas of the county, Spanish, Polish, Creole, or some other language is spoken by many workers. You should make arrangements for such people by having someone who speaks the foreign language available to make the announcement. Another idea is to make a recording of the generic announcement in the foreign language and play the recording during an evacuation. As a final suggestion, the floor captains should be aware that foreign language speakers may not understand the evacuation notice and may need to be directed out of the building.

For those rare buildings that do not have an audible alarm or a public address system, the building occupants can be notified of an emergency via a "telephone tree." The telephone tree works when one person is assigned a list of three people to call and provide the emergency warning or update. Each of those three people has a list of three people to call as well. Basically, it's the "friend told a friend, who told a friend" scenario.

The telephone tree system works well when a public address system is not available and the event does not warrant the setting of the audible alarm. For example, let's imagine a shopping mall setting and a severe weather advisory has been issued by the National Weather Service. You wouldn't go to a pull station and set the alarm; in fact, because of the severe weather, you certainly don't want people evacuating the building. You may elect to make an announcement via the public address system, but that may cause unnecessary concern as well. The telephone tree is an effective mechanism to warn shopkeepers, who could in turn warn the store patrons without raising undue concern.

Floor Captains and Brigades

Floor captains can be very beneficial in the event of an emergency. Their purpose is to assist the employees and visitors to safely and efficiently evacuate the building. They may be trained in first aid and basic fire extinguishing techniques. Floor captains provide special attention to those with disabilities, and they ensure that everybody has left the building. In some businesses, both floor captains and a fire brigade may be established. The responsibilities of a fire brigade are greater and require more advanced training. OSHA defines a fire brigade as "an organized group of employees who are knowledgeable, trained, and skilled in at least basic firefighting operations." Fire brigades are used when the nature of a business exposes its employees to unusual danger and the delay in response from the local fire department would cause further risk. For example, a nuclear power plant may find it prudent to establish a fire brigade to commence the firefighting process prior to the fire department's arrival.

Fire brigades are not required by OSHA, but if you do elect to create a fire brigade, a few regulations do exist. OSHA suggests that pre-fire training should be conducted by the local fire department so the fire brigade will be familiar with the workplace and potential hazards. This meeting and training also fosters cooperation between the local fire department and the fire brigade so they will be able to work together in the event of an emergency. If you desire to create a fire brigade, consult with your local fire department and review OSHA's regulations at CFR 29 1910.156.

Shut Down Procedures

When evacuating a building, it's usually not a good idea for everybody to just leave. Certain equipment and supply lines could present a major problem if left operational. For instance, suppose you have a natural gas service line entering your building. This line fuels four pieces of equipment. If these lines are left open during an evacuation for a fire, the risk of the fire being further fueled by the natural gas exists. A small fire could quickly get out of control and endanger the lives of the firefighters and other rescue personnel. To solve this problem, assign knowledgeable personnel the responsibility to shut down lines and equipment that might interfere with recovery efforts prior to evacuating the building. This requires identification of potentially dangerous equipment, storage tanks, containers, supply lines, and valves.

> Don't fall into the trap of identifying only lines that carry flammable or combustible substances. Flooding due to a broken water service line could hamper business operations and therefore create an emergency event. Remember to locate all supply lines and valves, whether they be gas, diesel, propane, chemical, water, or other.

The location of storage tanks and containers used to hold flammable or toxic chemicals must be communicated to the fire chief. This information will play a definite role in how the fire department uses its personnel and how it will fight the fire. Attempting to explain the hazards and their locations without a floor plan is folly. To make the floor plan user-friendly, the following information should be prominently identified:

- Shut-off valves
- Shut-off switches
- Storage tanks and containers
- Supply lines
- Breaker panels

Even when a plan is available, in the haste and confusion of the emergency, important features could be difficult to locate. To solve this problem, the visual floor plan must be cross-referenced with a written location system. The location key (see Figure 6-1) is a good way to accomplish that. The location key is created in a table format. The critical points are identified in the leftmost column. The middle column identifies the location of the items. The location is derived from an easy-to-implement grid system. The grid system can be superimposed over the electronic version of the floor plan in a CAD system, or you can simply draw a grid over a hard copy of the floor plan. The grid system works on an x and y axis. The horizontal axis (x) would be divided into segments and labeled with ascending letters of the alphabet. The vertical axis (y) would also be divided into segments and labeled with ascending numbers. Much like a map, you can use this grid system to quickly locate important items. The location key is most helpful when the items are arranged alphabetically. The final column includes a space for comments. This location grid should be attached, manually or electronically, to the actual hard copy of the floor plan.

Item	Location	Comment
Breaker panel	A-4	
Chemical storage	K-2 / P-17	See MSDS / Inventory log
Natural gas shut-off valve	H-12	
Water shut-off valve	H-13	Valve key above valve

Figure 6-1. Location Key

At the actual shut-off locations, make certain the valve is clearly identified. A sign with letters no less than two inches in height should identify the type of shut off. Have you ever been confused by how to turn a valve off? Do you turn the valve clockwise, or is it counter-clockwise? It's not always apparent, so be sure to label (at the valve location) the correct directional rotation for on and off. It may be advantageous to color code the sign depending on its use. For instance, a red sign with white letters may be used for all flammable and combustible valve locations. A blue sign with white letters may represent water shut off valves, and so on. Finally, always be aware of special keys or wrenches that may be required to turn the valve. Valuable time will be wasted if you don't know where the key is located.

Because of the increase in chemical and biological threats, many building owners and managers have realized a need to shut down the building's heating, ventilating, and air conditioning system in the event of an emergency. Chemical or biological agents can be quickly distributed throughout the building via the HVAC system; therefore, it makes sense to create a single shut-off switch for the HVAC system. To protect the HVAC system at its weakest point, pay particular attention to the fresh air intakes. If left unprotected, dangerous substances could be introduced into to HVAC system via the air intakes. From one location, an entire building can be infested with chemical or biological agents. Fresh air intakes can be protected in a number of ways. Simply installing security cameras at the intake location may be a good deterrent. A better way to protect the intake is to limit access to it. This can be achieved by installing fencing around the area where the intake is located. When installing fencing remember that some chemicals such as pepper spray can be sprayed into air intakes from a distance. The installation of motion sensors within the parameters of the fence would serve as further protection. In a multi-story environment experts suggest moving the air intake up from ground level, perhaps as high as four or five stories from the ground level.

Assisting Those with Special Needs

The orderly evacuation of all occupants, including those with disabilities, can be quite a challenge. Not an impossibility, but a challenge nevertheless. The Americans with Disabilities Act (ADA) offers many excellent suggestions for assisting those with disabilities during evacuation of a building. Those with hearing, vision, mental, or mobility issues need to be identified, and their egress from the building must be addressed prior to an emergency event, not during it.

When developing specific evacuation procedures for those with special needs, typically two types of solutions are available—engineering and procedure-driven. Engineering solutions might include slings or controlled descent devices on the market that allow the disabled person to exit via the emergency egress stairwell. Procedure-driven solutions might include the development of a partner system or the creation of areas of refuge for those with disabilities to wait for assistance. Often a combination of both solutions will provide the best protection.

Even though you attempt to cover all of the possibilities, you cannot fully protect your employees unless you are aware of their limitations. People with permanent or long-term disabilities are relatively easy to identify and assist. The challenge occurs when dealing with people with temporary limitations such as pregnancy or those aided by crutches. Think also about the individual with asthma, who might be able to easily evacuate the building under normal circumstances but, because of the stress and presence of smoke, may have difficulty breathing. When developing your emergency plan and building evacuation procedures, give consideration to these situations.

Use of Elevators For Exit

During an emergency, elevators are not considered a permissible means of egress from the building. In most buildings, the elevator has been programmed to immediately descend to the bottom floor (unless a fire is present at the bottom floor) upon being triggered by the alarm. At this point, the elevator is rendered inoperable unless emergency response personnel activate it using a special key. Many people have wondered why elevators cannot be used. Some studies have shown that if elevators were used during evacuations, building occupants could leave the building up to 50 percent more quickly. In particular, those with disabilities could safely and efficiently evacuate, without impeding the progress of others.

Using elevators during an evacuation sounds like a good idea until you understand why they are off limits. Firefighters must have a way to get to the fire. Since building occupants are exiting the building and fire rescue personnel are entering, there exists an opportunity for interference. Use of the elevator by fire rescue personnel allows them to get to where the emergency is in a quick fashion. Imagine if the firefighters, with their heavy rescue gear, had to climb 70 flights of stairs! Its much easier for building occupants to walk down 70 flights of stairs than it is for fire rescue to walk up 70 flights of stairs.

Another problem with elevators is the increased potential for smoke inhalation. Elevators cars are not very airtight. Doorways at each floor allow smoke to enter the elevator shaft. Smoke inhalation is not as likely for fire rescue personnel because they carry self contained breathing apparatuses (SCBA). Building occupants attempting to use an elevator during evacuation might not make it out alive.

The fact remains though, elevators could speed the evacuation, especially for those with disabilities. The National Institute of Standards and Technology (NIST) has studied the feasibility of elevator use during an evacuation, and they have developed a series of recommendations. First, the enclosed lobby area at each floor and the elevator shaft would need to be pressurized so that both remain smoke free. Secondly, dual power systems would need to be installed for reliability. Lastly, elevator components would need to be sufficiently waterproof to prevent failure due to flooding of the shaft during firefighting operations. Given these requirements, there has not been a strong movement to build elevators that can be used by building occupants during an evacuation. It's also safe to imagine that if such elevators did exist, their use would not be limited to those with disabilities. A bottleneck could occur as occupants wait for the elevator to arrive.

Evacuation Assistance Devices

Those persons with mobility limitations will often find building evacuations to be a difficult and sometimes precarious event. Imagine someone attempting to navigate a wheelchair down a stairwell. Now, add hundreds of people to the scene and you can appreciate the difficulty with many evacuations. Even people with temporary disabilities will find evacuation difficult. To assist these individuals, a number of evacuation assistance devices have been developed. If your building is a multi-story structure, you should seriously investigate the evacuation assistance devices that are on the market.

Among the evacuation devices available are mechanical devices that allow a controlled descent in the stairwell. One such device allows the wheelchair user to roll onto a platform, secure themselves to the platform, and then descend to safety. The advantage of this device is the wheelchair user never has to be removed from the wheelchair. Other devices require the disabled person to be separated from the chair. This introduces an opportunity for injuries and may create an uncomfortable emotional separation for the person with mobility limitations. Another device requires the wheelchair user to transfer from their wheelchair to a transportation platform which travels on specially installed tracks down the stairs. The descent speed can be controlled by a braking mechanism. When considering different evacuation devices, consider the distance the person must travel, the person's confidence in the device, the impact on stairwell space, and the cost. It would be advisable to ask wheelchair users for their advice and comfort level prior to equipment selection. If they are uncomfortable with the device or they do not know how to operate it properly, the device will likely go unused. When a decision is made on evacuation devices, be sure to train disabled persons and others how to use it properly.

Other non-mechanical devices are available to assist disabled persons with an evacuation. Various straps and harnesses are available that assist in the evacuation. If you do not have any of these types of evacuation assistance devices and you are faced with manually assisting someone in a wheelchair, at least use a safe technique. For instance, two people can efficiently guide a wheelchair and its occupant. Start by positioning one person behind the wheelchair grasping the pushing grips. Next, tilt the chair backwards until balance is achieved. While keeping the balance, move the wheelchair forward toward the steps. As the chair moves down the steps, always remain one step above the chair. Keep your center of gravity low and let the back wheels gradually lower to the next step. The wheelchair occupant will be leaning back and they may try to resist this position by leaning forward. Remind them not to lean forward. Kindly ask them to remain in the chair, lean back, and let assistants control the chair's descent. Be sure to have a second person assist by positioning themselves in front of the wheelchair to keep the wheelchair from gaining too much speed during the decent. Both assistants will attempt to keep the descent as smooth as possible, but they should not lift the chair off of the ground. The weight of the disabled person and the wheelchair could easily injure the assistants. A third assistant may be needed at the rear of the wheelchair to share control of the push grips.

> If any evacuation devices are used, make sure that the location of the device is noted in the emergency plan and on the emergency action floor plan. When an evacuation drill is conducted, use the devices. The time to get comfortable with these devices is not during the excitement and confusion of a real emergency. The emergency action plan must also address training on proper evacuation device use, including safe lifting techniques.

Regardless of the techniques utilized, don't forget to use proper lifting techniques. In past evacuations, people have hurt backs, twisted knees, and pulled muscles because they did not use proper lifting techniques. Usually the people who are hurt during these evacuations are the big and strong men who believe they can lift someone on brute strength alone. Meanwhile the smaller person who uses good techniques will come out uninjured. Some of the basic lifting techniques include

- Bend your knees
- Keep your back straight
- Use your leg muscles to lift
- Hold the person close before lifting
- Don't lift from an awkward position
- Ask for help if it is needed

Areas of Refuge

The Americans with Disabilities Act (ADA) outlines specific requirements for fire-resistant spaces where persons unable to use stairs can call for and wait for evacuation assistance from emergency personnel. These areas are commonly known as "areas of refuge" and are often located at egress stair landings, but other areas that meet specific fire rating and smoke protection can be used. ADA regulations require that two-way communication devices be installed in these areas of refuge so those awaiting assistance can be identified.

The use of a refuge area should only be used when the disability seriously impedes the evacuation of the individual. For instance, a wheelchair may present problems because of its weight and size, and you may elect to bring that per-

> An excellent resource for ADA questions, ideas, and resources is the Job Accommodation Network, a service of the Office of Disability Employment Policy of the U.S. Department of Labor. They have an excellent web site at <www.jan.wvu.edu>. They can also be reached at the following numbers:
>
> 800-526-7234 & 800-ADA-WORK (V/TTY)
>
> 304-293-7186 (Local Line, V/TTY)

son to an area of refuge. On the other hand, a person with a hearing or visual disability should be evacuated with the other building occupants, using the stairwell and emergency exits. An area of refuge is not a place to bring all people with any disability.

The Staging Area

An orderly evacuation from a building during an emergency saves lives. Think for a moment about an evacuation. Upon notification, personnel will begin to file out of the building, following the appropriate exit signage and routes. Eventually, all of the building occupants, perhaps thousands of people, will be standing in the parking lot or some other location. Now imagine the fire truck or ambulance driving up to the building. Will they be able to access the site? Will people be endangered as they wander around the site? Will they interfere with the emergency responders? For these reasons and many more, a staging area outside of the building must be developed.

- Employees should be instructed to move away from the exit doors.
- Ample staging area space should be provided for employees.
- Exit doors should not empty to parking spaces or streets.
- Inclement weather such as rain, snow, or heat should be considered.
- Groups should be kept relatively small.

The last suggestion is important. A large congested group of people wandering the parking lot creates confusion, and it difficult to account for who has exited and building and who remains. One concern that must be addressed when evacuating the building is safety. Is it safer to stay inside? Is the exterior staging area more dangerous than remaining indoors? Inclement weather, biological attack, or nuclear accidents are examples that may make an evacuation dangerous. The decision is probably not one to make internally. Guidance from law enforcement agencies or rescue personnel should be sought. If you do elect to evacuate during inclement weather, is an alternate staging area available? If an earthquake or fire has occurred, is the staging area far enough from the building to keep the evacuees safe? These are the types of role-play questions you'll need to address in your safety plan.

Accounting for Employees and Visitors

When evacuating a building, employees should have a specific area that they go to and remain at until the decision is made to go back into the building or to leave the premises. This staging area serves a few purposes. First, the staging area keeps the crowd of building occupants away from the building and out of the way of emergency responders such as police, paramedics, and firefighters. Secondly, the staging area provides emergency team members with an organized way to account for employees and visitors. This directive is so important that the Occupational Safety and Health Act requires it. OSHA

29 CFR 1910.38(a)(2)(iii) states procedures must be in place to account for all employees after an emergency evacuation has been completed. Think about the process. Where will employees gather? Who will perform the roll call? If somebody is unaccounted for, do you have their cell phone or pager numbers? If somebody is missing, how will this information be related to the emergency rescue personnel? Who will give the "all clear" order? These are the types of questions that must be answered if you hope to conduct an orderly and safe evacuation.

> In the event of an actual emergency, who will give the command to evacuate? What type of emergency warrants an evacuation? Is a partial evacuation sufficient, or should the entire building be evacuated? These are the types of questions that must be considered when developing your emergency action plan. A chain of command must be established and an individual must take the lead. Time is precious and bad decisions can cost lives.

It's important that employees understand the reasons for the staging areas. If the building occupants simply exit the building and mingle in the parking lot, it becomes difficult to perform a roll call. When this happens, emergency rescue personnel are placed in harm's way because they are searching the building unnecessarily for people who are already outside of the building. Keep the groups in the staging area small, 20 people or less per group, if possible. Small groups are easier to manage. Someone should be assigned the group leader whose primary responsibilities would include keeping the group together, performing roll call, and serving as spokesperson to the emergency command post.

The Chain of Command

During an evacuation, order must be kept. If not done in an orderly fashion, building occupants can be injured, time can be wasted, emergency response personnel can be injured, and company liability will be increased. It's critical that a clear chain of command be established prior to any evacuations whether they be training drills or actual emergencies. Here are a few of the important roles that make up the chain of command and which must be filled if an effective evacuation is to be conducted:

- Fire Brigade: The fire brigade will perform basic first aid and extinguish small fires in the event of an actual emergency. In an industrial environment they may fight more advanced fires if properly trained and supplied. They may also assist the floor captains by directed building occupants to the available evacuation routes, but this is not their primary responsibility.
- Floor Captains: The floor captains ensure that all building occupants have recognized the alarm and quickly head for the evacuation routes. Floor captains may also assist those with disabilities or those who do not speak English. Floor captains may be placed at stairway and elevator locations to monitor their use. The will also check

restrooms, loud locations, and private offices for persons who may not have heard the alarm.

Group Leader: The group leader will gather employees in the pre-assigned location in the parking lot. They will take roll call and will call missing employees or visitors on cell phones or pagers. If contact cannot be made, they will relay this information to the runner.

Runner: The runner will serve as a liaison between the group leader and the command post leader. During the evacuation the runner visits each individual group and gets a status report from the group leader. The runner will then relay this information to the command post leader.

Command Post Leader: The command post leader collects employee status from each of the groups and compiles a list of unaccounted personnel. The command post leader would pass this information on to emergency response personnel. If the evacuation is a drill, the command post leader would announce the "all clear" and direct all employees back into the building. If an actual emergency occurred, the command post leader would consult emergency response personnel to make the decision to re-enter or leave the premises.

Because the group leader will be conducting a roll call, they must have a current list of employees and visitors. The list of employees is easy to compile. If the group is kept small, the list will be easy to keep current. It should include the employees' names and cell phone, pager, and home phone numbers. The list should be kept on a small wallet-sized laminated card. If your group includes fire brigade members, floor captains, or employees responsible for equipment shut down, they likely will not be exiting the building until they have completed their assigned responsibilities and may not be in the staging area when the roll call is conducted.

One of the most difficult aspects of the roll call is accounting for visitors such as vendors and customers. Most of the time these visitors will be in the presence of an employee. If an evacuation occurs, the visitor should follow the employee out of the building. The employee should instruct visitors to follow them. If the employee and their visitor are separated during the evacuation, make this known to the group leader as they are taking roll call. The names of visitors who are unaccounted for will be passed on to the command post leader. Any group that finds an unknown visitor in its midst should verify the person's name and the name of the person they were visiting. You should also instruct that person to stay put. It may be helpful to retrieve visitor logs from the reception area and use those for accounting for non-employees.

If you plan to conduct an evacuation drill, someone should be assigned as the drill coordinator prior to the drill. This person plans, conducts, and evaluates the drill. A best practice is to rotate this assignment so various members of the emergency action team will understand the position. It is always advisable to have trained alternates, and this is a great way to ensure that will happen.

A Special Note for Those Who Refuse to Participate

There is no question about it; evacuation drills take time. Since time is valuable and we have so little of it to spare, some employees may resist taking part in the emergency evacuation drills. How will you handle this difficult situation? After all, you can't physically remove someone from the building, especially during a drill. This issue must be resolved and resolved quickly. If certain employees are allowed to sit out the drill because they are too busy, soon other employees will do the same. Before long you'll lose control of the drill.

The best way to ensure a well-participated evacuation drill starts with the organization's leadership. Management should clearly state their endorsement of the evacuation drills and make it clear that all employees are expected to participate. Of course, this includes executive management. They must be visibly participating as well! Management should put some bite into this requirement as well. For instance, a process could be established whereby the name of any employee not participating in the drill will be forwarded to the vice president of Human Resources. When the organization's leadership takes a leadership role in emergency planning, the employees will follow their example.

Employers also have the right to require non-participating employees to sign a waiver, releasing the company from liability if they fail to evacuate. The legal credibility of such a waiver may not be ironclad, but at least the employee will understand the importance placed upon the drill. Additionally, the burden of the added paperwork may be more painful than participation in the drill. Some organizations have made the required paperwork for non-participation so long and involved that the employees spend less time in an evacuation than they would filling out the paperwork!

If you have employees who refuse to evacuate or if they just grumble about it, they likely don't realize the importance of the drills. This is where training and education can help. The employee may feel burdened by meetings or important conference calls. Explain to them that it's permissible to end a meeting because of the drill. Teach them to tactfully excuse themselves from conference calls. Train them how to halt a meeting with a client and help them evacuate. Sometimes they just need permission to do these things. Give them the permission they are looking for.

Getting Back to Work

After the drill has been completed, the building occupants will be required to re-enter the building in an orderly fashion. This may seem like an obvious and simple process, but it may not be as easy as you'd think. If your building is a large multi-story structure, the re-entry process can be especially difficult and frustrating. Most large buildings are not equipped to handle the entry of all occupants at the same time. For instance, the thousands of tenants in a large building such as the John Hancock building in Chicago realize that if everybody started work at 8:00 A.M., it could take an hour or longer to get to your office. The lines at the elevator would be so long you would need to show up at the building's lobby at 7:00 A.M. just to get to work on time. To compensate for the rush of workers starting each day, most tenants stagger their arrival times. This serves to alleviate the crowds. By requiring some employees to start work at 8:00 A.M., others at 8:15 A.M., yet others at 8:30 A.M., you effectively lessen the impact of the mass of people entering the building.

The same concept applies to building entry after an evacuation drill. When the drill is completed, employees will break from their staging areas and walk directly to the elevators. Imagine the long lines, the chaos, the frustration. The next time you attempt to practice an evacuation drill, you'll find many people unwilling to participate because of the time consumed in re-entry and loss of productivity. To solve this problem, two good options exist. First, you may elect to conduct an evacuation drill in smaller groups, perhaps two or three floors at a time. This makes the management of the groups and re-entry into the building much easier. You'll need to devise a way to alert the occupants of those floors being evacuated, without tripping the alarm and triggering an entire building evacuation. If your alarm system does not allow for a selective siren or message, you may decide to use your floor captains and a bullhorn. If you use this option, try to simulate the actual sound of the regular building alarm. A simple tape recording of the alarm played over the bullhorn may suffice. Another option is to purchase a small alarm horn or siren at an electronic store. Your desire is to condition the building occupants to recognize the alarm and immediately begin evacuation.

> Evacuation of educational facilities should take into account the hazards that exist and the comprehension level of the students. Because of the young age of the majority of the occupants in an educational facility, the administration must provide an increased level of direction and hands-on assistance when evacuations are required. Questions to ask include:
>
> How will we communicate with parents?
>
> How will we account for students who are picked up by parents?
>
> Can transportation be brought in to take students to a safe location?

The second option is to conduct a complete building evacuation, but instead of allowing the building occupants to re-enter the building all at once, you will release them in small groups. This is best done by releasing various staging areas in a staggered fashion. Groups would be released based on their location in the building and which elevator banks were available. All elevators, including cargo elevators should be utilized during the drill. If an elevator or escalator is out of service, you might be better served by postponing the evacuation drill until you have all systems operational.

Evacuation captains and assistants will need to have communication via two-way radio to ensure a smooth release of building occupants. Some additional planning will be necessary, but this option more closely represents an actual emergency situation.

When an actual emergency occurs that requires an evacuation, law enforcement personnel may want to interview some of the building occupants. Employees who were in the area of concern or who witnessed a problem should be identified and made available to the investigating agencies. A simple process can be created to ensure that this happens. For instance, the involved employees should have evacuated the building and assembled at their assigned staging area. Involved employees should check in with their staging area group leader and inform them of what they witnessed. The group leader would share this information with the runner, who would in turn deliver the information to the command post leader. This structure will allow important information to be relayed to a central location in a quick and efficient manner.

After the Drill: Evaluating Performance

The emergency action plan should be evaluated on an annual basis at a minimum. No plan is ever perfect. Even if it were, people and processes change. A plan that sits on a shelf until an emergency occurs will be an outdated plan. In addition to the annual review, time should be taken to review the evacuation procedures after a drill is performed. The responsibility of auditing the emergency action plan should be assigned to a specific employee. Usually this will be a member of the emergency management team. That's not wrong, but consider assigning the audit to an employee who is knowledgeable about safety and emergency planning, but who did not have a role in developing the emergency action plan. They will often bring a fresh insight and new ideas to the table.

If the emergency evacuation was a drill, let the group know how they did. Was the clocked time within the established benchmarks? Did they do better or worse than the last time the drill was practiced? If it was better, share that success with the employees. They need to know that the effort they have expended has been worthwhile. If the timing was substantially worse, ask the group why they think the performance was poor. They may be able to provide valuable insight into problems and the corrective measures that should be taken. If you keep your building occupants apprised of the evacuation results, they will take an interest. They will want to improve their evacuation time the next time the drill is conducted. Most importantly, in the event of an actual emergency, the building occupants will have the confidence to evacuate without resorting to pushing, running, or other inconsiderate and dangerous behavior.

One of the areas that frequently becomes outdated in most emergency action plans is important support personnel data including names and phone numbers. This may include employees within the organization and contractors or governmental agencies outside of the organization. The auditor could confirm that individuals listed are still in the organization and still have the responsibilities that caused them to be included in the plan. Home, cell phone, and work telephone numbers should be verified. In addition to names and numbers of support personnel, how would you answer the following questions:

- Does the organization have a formal plan for training its employees in emergency procedures?
- Is training being completed?
- Has your organization met with local community and government agencies and briefed them on your plan?
- Is the organization's insurance coverage adequate?
- Do you have updated photos or videotape of the facility and its assets?
- Does the plan reflect changes in the building layout?
- Does the plan reflect changes in policies and procedures?
- Does the plan reflect lessons learned from the drills and actual emergency events?

Basic First Aid and Medical Care

During the course of a typical workday, employees will experience cuts, bruises, strains, and other injuries and inconveniences. When these events occur, basic first aid should be administered. When a workplace emergency occurs, injuries are likely to occur. History has shown us that many of the people who are injured can be assisted with basic first aid. Often during an emergency situation, emergency responders will be delayed and critical first aid will not be readily available unless you are able to provide it in-house. In this case, the first aid responder will provide the initial assistance that may help the victim survive until trained emergency medical technicians arrive. Providing first aid need not be an expensive proposition. You are not required to hire licensed physicians or nurses. Thanks to "Good Samaritan" laws enacted in most states, you do not need not be overly concerned with the added liability of providing first aid. Basic first aid, cardiopulmonary resuscitation (CPR), automated external defibrillator use, and emergency oxygen administration are some of the first procedures that are available to most organizations.

The First Aid Kit

A first aid kit is an essential element of emergency medical response. The contents of your first aid kit should treat the types of injuries that might occur in your workplace and suit the expertise of the first responders providing initial care and relief. Cuts and burns often represent the greatest exposure in the workplace, so ointments and bandages are a minimum requirement. Other suggestions for the first aid kit might include

- Adhesive tape
- Adhesive strips (¾" by 3" and 1" by 3")
- Ammonia inhalants
- Antacid tablets
- Antiseptic ointment or wipes
- Aspirin tablets
- Blanket
- Biohazard bags
- Cold pack
- Cotton tip applicators
- Eyewash
- Fingertip bandages
- Gauze pads and roller gauze (assorted sizes)
- Antiseptic hand cleaner
- Ibuprofen tablets
- Knuckle bandages
- Latex gloves
- Non-aspirin tablets
- Notepad and pen (actions and inventory)
- Plastic bags
- Scissors and tweezers
- Small flashlight and extra batteries
- Triangular bandage

> Remember to include in your kit a log of incidents and a log of first aid contents used. These logs will serve a two-fold purpose. First, they will ensure the contents are reordered as they become depleted. An empty first aid kit does not benefit anyone! Secondly, the logs will provide you with a process to spot trends and take proactive steps to protect employees and visitors while in your buildings.

Depending on the environmental exposure and the in-house expertise, these additional first aid supplies might be suggested:

- Activated Charcoal (use only if instructed by Poison Control Center)
- Assorted splints
- Blood pressure kit
- Burn relief spray
- CPR mask with oxygen port
- Defibrillator
- Insect bite relief spray
- Insta-Glucose
- Nitrile gloves
- Stethoscope
- Syrup of Ipecac (use only if instructed by Poison Control Center)

> First aid kits can be purchased from a variety of sources. Companies such as PS Plus <www.publicspaceplus.com> and Colonial Industrial Products <www.cip-safety.com> provide excellent "ready-to-go" first aid kits for a variety of needs and exposures.

When assembling a first aid kit, create a first aid station as well. This station might be located in the shop area or in a break room and would consist of the first aid kit, first aid posters, and a log of incidents and contents used. The station need not be expensive; it could be as simple as a plywood board painted and attached to the wall. The employees should be trained in locating the first aid station, administering proper first aid care, and reporting injuries. In a large building, multiple first aid stations are recommended. It would not make sense for an employee working on the second floor of the building to be required to travel to the fifth floor to receive assistance.

Burn Victims

In the event of an emergency, especially one that involves burns, proper care must be taken so as not to harm the victim further. Burns can be extremely painful, causing shock, infection, and even death. Before administering the proper first aid, you must be able to recognize the type of burn to be treated.

There are four types of burns:

- Thermal. These burns are caused by fire, hot objects or liquids, and gases, or by nuclear blast.
- Electrical. These burns are caused by electrical current or lightning strikes.
- Chemical. These burns are caused by contact with wet or dry chemicals.
- Laser burns. These burns are caused by the concentrated heat emanating from a laser device.

Depending on the severity of the burn, one of three classifications will be assigned:

- First-degree burns. These are relatively minor burns in which only the outer layer of skin is burned. The skin is usually red and some swelling and pain may occur. Usually can be treated by first aid.
- Second-degree burns. These are burns in which the first layer of skin has been burned through and the second layer of skin is also burned. In a second degree burn, the skin reddens and blisters develop. Severe pain and swelling also occur. This burn should be examined and treated by medical professionals.
- Third-degree burns. This is the most serious type of burn because it involves all layers of the skin. Fat, nerves, muscles, and even bones may be affected. These burns should receive prompt emergency

medical attention. Delay in treatment may result in further damage, shock, and infection.

The scope of this book does not allow an in-depth discussion of the four types of burns or the three degrees of burns; however, a few points are worth making.

When treating a burn victim, first attempt to determine the cause of the burn. If the burn is the result of electrical contact, do not touch the victim until you are sure they have been removed from the electrical current. Failure to do so will result in you being shocked and burned as well.

When treating minor burns follow these steps:
- If the skin is not broken, run cool water over the burn for several minutes.
- Cover the burn with a sterile bandage.
- Administer acetaminophen to relieve any swelling or pain.

In the event of a major burn, seek emergency treatment immediately. Until the arrival of trained medical personnel, follow these preliminary steps:
- Remove the person from the source of the burn (i.e., fire, chemical, electrical current)
- Ensure that the person is breathing. If they are not, begin mouth-to-mouth resuscitation immediately.
- Remove all smoldering clothing to stop further burning.
- Do not apply any ointments, creams, or ice on the burned area or break blisters.
- Stay with the victim and reassure them that they will be okay and that help is on the way.

Chemical Exposure

In the event of chemical exposure, immediately remove the victim from the area of exposure; however, you must do this without exposing yourself. Begin by contacting 911 or your local emergency management agency. When attempting to enter the exposure area, you may need gloves, a respirator, or even SCBA gear. Before entering the exposure area, determine what chemical has been released. You may be able to do this by observing opened canisters, ruptured hoses, or other telltale signs. When the chemical is identified, immediately retrieve the Material Safety Data Sheet (MSDS) for that chemical and find the suggested treatment and warnings.

If you do not have the MSDS information available, visit an onsite MSDS database (available at <www.msdssearch.com>) and search for the chemical name. Because of the lack of time or an Internet connection, be prepared to describe the chemical, its name, smell, color, and the effects it is having on the victim. Of course, with any chemical exposure or burn, time is of the essence and should not be wasted. Every second counts.

Broken Bones

Broken bones are a common byproduct of emergency events. Broken bones include minor fractures or multiple breaks that encourage shock and nerve damage. When examining a victim for broken bones, look for the obvious signs such as deformity, protruding bones, or bleeding. Less obvious signs include swelling, tenderness, pain, and bruising at the location of injury. Do not ask the victim to move the injured limb because movement of the fractured bone could cut through muscle, blood vessels, and nerves, further complicating the situation.

Professional medical attention must be given to the victim of bone fractures; however, in some scenarios, medical attention may not be readily available. If this is the case, fashion a splint to immobilize the fracture. Again, don't move the limb unnecessarily, and if you are not sure if the injury is a fracture or a sprain, assume the worst-case scenario and proceed accordingly.

Unless the victim is in immediate and life threatening danger, resist the temptation to move a victim with neck or back injuries because this could also cause further long-term damage.

Defibrillator

According to the U.S. Department of Labor, sudden cardiac arrest occurs at least 1,000 times a day and accounts for 75 percent of all workplace deaths. Advocates of Automated External Defibrillators (AED) claim that tens of thousands of lives can be saved each year by installing AEDs in public places such as airports, theaters, sporting arenas, shopping malls, and other places where people congregate.

AEDs are used to restore a normal heartbeat in an individual who has experienced ventricular fibrillation, more commonly known as sudden cardiac arrest. Cardiac arrest is brought about by sudden chaotic and abnormal electrical activity of the heart. Because the heart is able to pump based on electrical impulses, the abnormal electrical activity causes the heart to quiver in an uncontrollable fashion resulting in little or no blood being pumped from the heart. The victim of sudden cardiac arrest will quickly lose consciousness, and unless the condition is reversed, death follows in a matter of minutes. The AED provides a controlled electrical shock to the heart that helps to restore a normal rhythm.

When administering assistance using an AED, you'll also need to know CPR because these two methods usually go hand-in-hand. Prior to commencing the use of an AED unit, the administrator must evaluate the victim to determine need. The AED does not help someone who is choking and needs the Heimlich maneuver. For obvious reasons, the AED administrator must receive training. Using an AED device is not difficult. You do not need to be a licensed physician or paramedic. An in-depth knowledge of the medical field is not necessary; however, you can't expect an employee to arrive on the scene of an emergency and hook someone up to the device and save their life. How easy is an AED to use? If trained properly, even "sixth-grade school children have learned to use AEDs," reports the *Journal of the American Heart Association*. Training on proper use of an AED device is just the beginning for the people designated as first aid responder. The American Heart Association recommends five specific areas of training that should be required for first aid responders:

- Cardiopulmonary Resuscitation (CPR)
  CPR in conjunction with AED has proven to be an important step in treating a person who experiences sudden cardiac arrest. Not every emergency needs AED; sometimes only CPR is needed.

- Automated External Defibrillators (AED)
  AEDs are used to restore a normal heartbeat in an individual who has experienced sudden cardiac arrest. Sudden cardiac arrest is usually caused by ventricular fibrillation (VR). When ventricular fibrillation occurs, the heart ceases to pump in a normal rhythmic fashion, thus preventing the heart from pumping blood properly.

- Emergency Oxygen Administration
  A lack of oxygen in the victim's bloodstream can lessen the effects of an AED device, cause brain damage, and even cause death. Emergency oxygen can provide relief until emergency medical technicians arrive on the scene.

- Bloodborne Pathogens (BBP)
  Certain diseases such as Hepatitis and AIDS are carried in the bloodstream and can be transmitted to others, including first aid responders. A first aid responder must understand how these diseases are transmitted and how they can protect themselves from these diseases. The Occupational Safety and Health Act requires BBP training for anyone providing first aid.

Not every medical emergency requires AED or even CPR. Often first aid simply means the application of ointments or bandages. First aid responders should be properly trained in these techniques.

Prior to deciding to install AEDs in your place of business, here are a few items you'll need to consider:

- Decide how many AEDs will you install.
- Research where will they be located.

- Research all federal, state, and local regulations regarding AED usage.
- Create a process to keep informed of changes to regulations involving AED usage.
- Obtain a prescription from a state licensed physician as required for AED installation.
- Coordinate your AED program with your local emergency medical service.
- Develop training programs for those responsible for using AED.
- Develop a system to provide retraining and document compliance for "Good Samaritan" requirements.

When considering how many AED units to install and where to place them, consider this: the *New England Journal of Medicine* recommends that AEDs should be located within a three-minute "drop-to-shock" window. In other words, the shorter the time frame between the "drop" (the heart attack) and the "shock" (the AED application), the better the chance for survival. For each minute after the heart attack, the chance of recovery is reduced by ten percent. After ten minutes, the chance of survival, even with an AED shock, is extremely low.

> The American Heart Association has identified four links that make up the "Chain of Survival" for those experiencing sudden cardiac arrest:
>
> Early access (phone 911)
>
> Early CPR
>
> Early defibrillation
>
> Early advanced care (paramedic/physician)

Your emergency procedures manual should specifically address the AED device and its proper usage. Be sure to specify the names, locations, and phone numbers of those who have been trained in proper use of the AED device. This plan needs to be evaluated at least annually and revised as trained employees come or go. One final and very important feature of any emergency procedures manual is a floor plan that identifies exits, fire extinguishers, first aid stations, eye wash stations, and AED devices.

CPR

Cardiopulmonary Resuscitation (CPR) is a manual technique that keeps oxygenated blood flowing to the brain and heart when a person is experiencing sudden cardiac arrest. CPR bridges the gap between cardiac arrest and the time when advanced treatment, such as defibrillation, can be administered. Studies have proven that the survival rate of sudden cardiac arrest victims improves dramatically the sooner they receive CPR after the onset of cardiac or respiratory arrest. Further studies have shown that the administration of

CPR, which can be done quickly, bridges the gap between administration of automated external defibrillation, which usually is not readily available.

CPR is not difficult to perform, but proper training and certification should be completed before an attempt is made to administer it. Organizations such as the Red Cross (<www.redcross.org>) and the American Heart Association (<http://www.americanheart.org>) provide regular local CPR classes. CPR training is not required in most businesses, but a few exceptions do exist. For instance the logging industry, commercial diving operations, and rescue personnel assigned to permit-required confined spaces all require CPR training and certification.

Summary

One of the more difficult assignments for the emergency manager concerns the evacuation of people from a building. This is true whether your building is a large high-rise or a single-story office building. The concept of evacuation seems simple enough, but during an actual emergency event, people often become disoriented and confused. For this reason regular evacuation drills are necessary. Unfortunately, in many scenarios, full-scale evacuation of a building is simply not practical. In these cases partial evacuation or even tabletop exercises serve as a good alternative.

When considering the evacuation of a building, special consideration must be given to persons with disabilities. People with hearing limitations may not be able to hear an alarm. People with visual limitations may not be able to find their way out of the building via safe passageway. Employees and visitors with certain mobility limitations may find it difficult to leave the building, especially if the elevator is not available. Although these issues may make for a more difficult evacuation, they do not make it impossible. You have a responsibility to provide a safe evacuation for all employees including employees with disabilities.

7

Post-Event Restoration

After the Event

Often an emergency event will occur after business hours or when emergency team members are otherwise away from the office. After the emergency event, employees will be evacuated, roads may be closed, and access to "ground zero" may be limited. If this does occur, how will you and other emergency team members be able to reach your buildings? Law enforcement officers, or even the National Guard, will restrict all but the most critical traffic in and out of the area. This serves a number of purposes. First and perhaps most importantly, many of the access roads will be unsafe. In an effort to protect the public, road travel will be temporarily unavailable. Secondly, roads that are available must be reserved for emergency response personnel. Travel on these roads will be slow if trees are blocking the road or if earthquakes have caused the loss of pavement. The slow travel could be further exacerbated when emergency response personnel are slowed down by non-essential vehicular traffic. A third reason access is limited is because of the increased possibility of looting and vandalism. After an emergency occurs, offices normally locked and guarded are now unprotected. Doors and windows have been broken and are open. Alarms are not functioning as a result of power loss. Police forces are spread thin. The opportunity for a thief to strike is at its highest. Many businesses have lost computers, furniture, files, food, tools, and other valuable assets to theft following an emergency.

Given these reasons, it is understandable why free access after an emergency will likely be unavailable. Law enforcement and rescue personnel have a duty to protect citizens, and they take their job very seriously! Your emergency response team will only be allowed access if you communicate with these agencies prior to an emergency. Have you met with local law enforcement and rescue personnel? Have you introduced your emergency response team? Have you discussed your emergency action plan? If you take these actions and develop a relationship with the local agencies, they will be more apt to grant access to your emergency response team.

These agencies may permit entry to a limited access zone, based on the critical nature of your business. For instance, utility workers will usually be granted greater access rights than a shopkeeper. At the initial meeting with the local law enforcement agency, be prepared to make your case for immediate access to your site. You will need to convince these agencies of the importance of allowing you to respond to your site. This should not be based on a desire to get to your site; rather, it should be based on a need. Is your business critical to community-wide disaster recovery? Does your business play a support role? Are employees on site? Can your response efforts decrease the need for police protection and allow them to work in other areas? Whatever your reasons for needing site access, be prepared to share this with local law enforcement and rescue personnel prior to an emergency.

These agencies may propose that certain identification be carried by emergency response team members. Perhaps a magnetic sign attached to the vehicle or a window hanger identifying the occupant of the vehicle as an emergency response team member would be sufficient. One word of warning, though, magnetic signs and other temporary forms of identification can be stolen. The holder of such identification now has unlimited access into a controlled zone, so many agencies prefer other means of identification. In many areas, law enforcement officers will have a list of emergency responders, and if your name is not on the list, you're not getting in. Find out if such a list is used and what is required to have the names of your team members added. Remember, the list of emergency team members will change over time and you must contact these agencies to revise their list.

When members of your emergency response team are called to action, they must be prepared mentally and physically for what they may find. If your team is an early responder, a group that responds within an hour after the emergency event, it is unlikely that team members will be able to predict the severity of what lies ahead. Water service could be disrupted. Heating or cooling may be nonexistent. Floodwater may impact their efforts. Lack of electric service will hamper recovery efforts. If the area has been hit hard by the emergency, food and water will be scarce. Finally, transportation out of the area may be difficult. Many an emergency team member has been stranded in an inhospitable environment without food, water, or a way to quickly get back to civilization. The emotional and physical trauma can cause early responders to feel hopeless and powerless. In severe cases, the responders may even shut down, physically and emotionally. This phenomenon is magnified with a lack of food, water, or sleep. Finally, the experience reaches a crescendo after the initial adrenalin rush has passed and the realities of restoration become evident. Understand the physical and emotional swings during a restoration effort and address them proactively.

A list of provisions must be provided to emergency team members and responders prior to an emergency. The list should be part of the emergency action plan, and friendly reminders should be given to team members on an annual basis at minimum. A list of personal inventory items can be found on pages 176—177.

Members of the emergency response team or employees asked to assist in the recovery effort should be accounted for and tracked. Essential information such as telephone numbers, home addresses, vehicles, location assignments, skills, and tools should be recorded. A sample Contractor Availability Form can be found in the appendix. This form can be an invaluable tool when trying to locate key employees post-emergency. After an emergency event, maintenance employees can be invaluable; however, they might not be members of your emergency response team. You need a process to quickly see who is available and what tools or skills they possess. The emergency responder identification form can do that. Additionally, this form can help facilitate relief efforts and ensure that all facilities are visited and audited after an emergency event.

> Be aware of subtle changes in your team members. If fatigue is setting in, make arrangements for rest and relief. Quite often more injuries occur during the grind of restoration than during the emergency event itself. If team members have experienced loss of a family member or personal property, they should be excused from the restoration efforts to care for the needs of their families.

Assessing the Damage

Now that you have arrived on the scene, you should immediately begin taking steps to assess the damage to your building and the surrounding property. Depending on the nature of the emergency event and the resulting damage, you may elect to bring in specialized help for this assessment. If you have any doubts about the structural integrity of the building, do not enter it! The appendix contains sample building assessments (General Building Issues—Annual Audit) that may be used to assess the buildings condition. These audits provide an excellent starting point for post-emergency recovery; however, they should be revised to incorporate the specifics of your property and site.

> When entering a building for an assessment, make certain you have the proper personal protective equipment available and in use. This might include hardhat, steel-toed shoes, long-sleeved shirt, nuisance dust mask, respirator, and safety glasses. Employees assigned PPE must be properly trained in its use, and with many items such as respirators, the gear must be fit tested.

Insurance and Salvage Decisions

As you walk through your property after an emergency event, you'll need to document any property damage that may have occurred. If you have an insurance policy on the building and its contents, this documentation will be essential if you hope to be reimbursed for your loss. If you are not familiar with the insurance coverage of your property, immediately contact your risk management department or insurance provider to make certain you are following correct procedures for documentation. If your organization is self-insured, you are not exempt from the documentation process. You will still need to document the damage if you hope to take a tax write off or take advantage of federal or state aid that may be available.

> Relying on the copy of the building assessment in your emergency action plan or on your computer's hard disk might be risky. If electrical service were compromised, you would be unable to use a photocopier or a printer. For this reason, you should have a number of preprinted assessment forms ready to go. A common practice is to covert these assessment forms to an electronic document accessible on a personal digital assistant. That's a great idea; just don't forget your battery charger and a source of power to operate the charger!

Some type of visual recording device should be used when performing an assessment of your building and the associated property. If possible, a video camera should be used to record the damage. Continuous video gives a more accurate and complete description of what has occurred than photographs. Remember, insurance adjusters and government agencies providing aid will be reviewing this documentation. If a video camera is not available, a digital or film camera may be used. The date and time stamp should be included when filming or taking pictures. If you do use a digital camera or video recorder, remember the limitations. How many pictures can the digital camera hold before its memory is used up? How long will the battery last? Do you have a means to recharge the battery once it is depleted? In areas where loss of electrical power has occurred, many emergency responders prefer to use simple disposable cameras. Others prefer to use instant-print cameras because they can write notes on the back of the picture while the thought is fresh in their minds.

Keep salvageable property separated from "damaged-beyond-repair" property. The damaged property should be kept until an insurance adjuster has had the opportunity to evaluate it. If the damaged goods are a hindrance to recovery and exposure to the elements will not further harm them, establish a temporary staging area on site and move those items to that location. Prior to moving these items, document their condition and where they came from. If salvaged or damaged items are sold or otherwise removed from the property, be certain to collect a signed inventory list from the party removing the goods.

After a particularly devastating disaster, insurance response may not be immediate. When Hurricane Andrew struck south Florida, many people who were affected said that it took weeks, and in some cases months, for the insurance companies to meet with them. Some insurance adjusters were unable to find their clients because the buildings were no longer standing and they had no alternate means of communication due to the loss of telecommunications. If a major disaster occurs and your building no longer exists or is hard to identify, build a large sign out of plywood and paint the address where you can be located, the company name, and your insurance provider on the board.

The Restoration Phase

At some point after victims have been aided and the damage has been assessed the actual restoration phase must begin. During this time, if unprepared, confusion, exhaustion, and often depression can set in. Having a detailed plan of attack will decrease the downtime caused by the emergency event. A few words of warning are in order, though. The task of rebuilding often seems monumental, even impossible. If the emergency event was a regional one, the ability to get quick-responding local help may be impossible. The pressure of getting the business back on line may be stressful as well. The restoration phase may require long hours supplemented by little sleep and hard work rewarded with sub-par food. You may have serious family concerns but little family interaction. These are the very real issues and emotions that are likely to be experienced by those supporting the restoration phase.

The restoration phase is not all pain and misery. A spirit of unity can overtake a previously apathetic community. You'll witness true human compassion being expressed on a regular basis. The ability to recover from adversity gracefully is uniquely human. It's a side we don't always see, or perhaps just don't notice. Sometimes the worst events bring out the best in people.

Logistics

One of the most difficult aspects of emergency recovery is logistics. Why logistics—or better yet, what is logistics? Logistics can be defined as the movement of equipment, supplies, and personnel to locations where they are needed most. This movement becomes difficult when the access roads have been damaged. It becomes complicated to achieve when telecommunications are down. It's nearly impossible when your employees and contractors are literally digging themselves out of the ruins. Even with all these issues satisfied, the rush to get back to business puts extra pressure on the restoration team. Things can't happen quickly enough. The problem with logistics is made more difficult when you have multiple damaged properties in your portfolio. You find yourself juggling conflicting priorities, schedules, and demands. Some of the logistical issues we will discuss in this chapter include

- Telecommunication
- Food provision
- Hotel accommodations
- Shower facilities
- Medical support
- Cash advances
- Continuation of salary
- Safety and security

One of the best solutions for logistic difficulties is to have an inventory of the basic necessities waiting and ready to go prior to an emergency event. Understanding that this inventory could become a casualty of the emergency (i.e., your inventory of food and water may be destroyed during an earthquake), it would be wise to spread your supplies over a few locations, if at all possible. If you are a member of the emergency response team or are counted upon as a first responder, you should be self-sufficient for at least 24 hours, preferably for 72 hours. Initially, the emergency responders have the responsibility to provide the basic necessities for their personal consumption. By having 72 hours worth of supplies, they can survive and be modestly comfortable until formal provisions can be initiated. Such provisions might include

Clothing
- Three changes of clothes
- Rain jacket
- Cold weather jacket
- Safety goggles
- Hard hat
- Sun glasses
- Two pairs of work shoes (steel toed preferred)
- Spare pair of glasses

Healthcare and toiletries
- Soap
- Shampoo
- Wash cloth and towel
- Razor and shaving cream
- Toothpaste and toothbrush
- Personal hygiene products
- Deodorant
- Toilet paper
- Sun block
- Insect repellant
- Contact lens solution and spare lenses

Miscellaneous
- NOAA weather radio
- Alarm clock
- Blankets
- Sleeping bag or cot
- Batteries
- Flashlight
- Pen or pencils
- Notepad and audit forms

- Camera (digital or film)
- Video camera
- Cell phone with automobile battery charger
- Driver's license
- Credit card
- Cash
- Plastic container to keep items dry

Medicines
- Prescription medicines
- Antacids
- Athlete's foot spray
- Pain relief cream
- Aspirin and ibuprofen
- First aid kit

Food and Water
- Meals Ready to Eat (MRE)
- Snack bars
- Fruit
- Trail mix or nuts
- Bottled water (drinking and cleaning)
- Instant coffee
- Cooler with ice and ice pack
- Propane stove (optional)
- Propane fuel tank (extra)

This list is not all-inclusive and, depending on the emergency, not all items will be necessary. The list is just long enough to provide a level of comfort without burdening the emergency personnel with a tractor trailer full of supplies. Remember, this inventory is for their consumption only. Encouraging emergency team members and first responders to be responsible for their 24–72 hour inventory will allow them to be better able to respond prior to the restoration of the infrastructure.

> Mr. John Lee, an emergency planning consultant based in North Carolina, provides following recollection of events after Hurricane Andrew:
>
> During Hurricane Andrew in August of 1992, I was supporting a large corporate building that housed nearly 2,000 employees and the corporate computer mainframe. We felt comfortable that we had planned for all contingencies. We had generator and uninterruptible power supply for the computer mainframe system. We secured the air handler units on the roof. There was enough food and potable water to feed a skeleton staff for several days. We knew there would be 300–400 employees in the building for several days just to keep communications up. I felt comfortable that we were prepared for the storm. We had a good plan, and the employees were properly trained to provide support after the storm. The building faired the storm well—just a few minor roof leaks. All emergency power systems were working and the computer mainframe did not miss a beat. The one item we did not plan for turned into a major inconvenience. We were unable to flush the high-pressure flush valve commodes. Unlike the ones most of us have at home, a bucket of water doesn't work. The municipal water company lost power for a few days and was unable to pump. The moral of the story: No matter how much planning you do, something unexpected will always crop up and present problems. Learn from your past and make adjustments as necessary. I don't write a plan without identifying Port-o-Let vendors!

The Command Center

The command center is an integral component of any emergency recovery plan. There must be a central location where the management of the recovery efforts can be performed without interruption. Even in an evacuation drill, a command center should be established as the focal point of communication.

Whether you elect to locate the command center at the damaged property or off-site typically depends on a number of factors. For instance, after an actual emergency that requires restoration, the command center must be equipped with phones, fax machines, computers, and other communication resources. If electrical service has not been restored and emergency generator service is not available, the command center should be located elsewhere. Having said that, in most cases the command center would better be served at the location where restoration is being done. In scenarios where you have multiple properties that have been damaged, a central command center should be established and then backed up by individual on-site command centers.

What happens at a command center? Again, the key word is communication. The command center serves as the "brain center" of the recovery effort. Contractors, employees, media, government agencies, insurance providers, or anyone else with an interest in the property needs to have a place to go to ask questions and get answers. Because of the tremendous amount of coordination that a restoration effort requires, the command center is a must. In addition to communication, the command center may be responsible for

- Personal protective equipment
- Meals and beverages
- Property security
- Accommodation arrangements
- Tool checkout
- Contractor sign-in
- Media relations
- Government agency liaison
- First-aid

Because the command center will become the focal point of communication, keeping up with the questions, answers, and promises made will become overwhelming if left to memory. All correspondence should be documented using a correspondence log. (Figure 7-1) The correspondence log works much like a telephone log, but it documents all conversation, whether by telephone, e-mail, verbal, or fax. The correspondence log tracks conversations with vendors, suppliers, emergency management teams, members of the media, or any other entity. Copies of press releases, contractor insurance certifications, PPE checkout, deliveries, and any other important issues can be documented in this book as well. So many things will be happening and so many promises will be made, that it becomes very difficult to manage an organized emergency restoration effort without the log.

The Emergency Account Number

We've all witnessed the devastating results of emergency events. Whether the emergency be manmade or natural, the results can be very costly. Most organizations understand the potential costs associated with these events and they attempt to share this risk by purchasing property insurance. In addition to insurance providers, certain governmental agencies may be able to provide financial relief after an emergency.

The insurance providers and government agencies may provide financial relief, but they do have rules. They will require certain documentation to prove that expenses were used for emergency recovery efforts. You must ensure that recovery costs are not commingled with standard operating expenses. The funds must be used to repair damage as a result of the emergency, and not to buy new artwork for the corporate office. Each expense will be examined by auditors and tested for validity. Additionally, the recovery expense documentation must satisfy your internal auditors. Speaking with your internal auditors and allowing them to review your current accounting procedures would be a good starting point. They can serve as a liaison between your organization and the insurance provider. Additionally, in the event of an emergency the auditors should be relied upon to provide specific guidelines for what can be charged to the emergency account number and what cannot.

Correspondence Log

Date: _____ **Time:** _____

Caller	Issue	Name	Office Phone	Cell Phone	Action Required	Assigned To
☐ Vendor ☐ Contractor ☐ Employee ☐ Police ☐ Fire / Paramedic ☐ Insurance Agent ☐ Other _____					☐ Yes ☐ No	

Date: _____ **Time:** _____

Caller	Issue	Name	Office Phone	Cell Phone	Action Required	Assigned To
☐ Vendor ☐ Contractor ☐ Employee ☐ Police ☐ Fire / Paramedic ☐ Insurance Agent ☐ Other _____					☐ Yes ☐ No	

Date: _____ **Time:** _____

Caller	Issue	Name	Office Phone	Cell Phone	Action Required	Assigned To
☐ Vendor ☐ Contractor ☐ Employee ☐ Police ☐ Fire / Paramedic ☐ Insurance Agent ☐ Other _____					☐ Yes ☐ No	

Figure 7-1. Correspondence Log

It is one thing to provide an emergency account number; it is quite another to get employees to use it. If the paperwork is too complicated, the employees will not fill it out and the company runs the risk of not being compensated for its restoration-related expenses. All employees need to understand the importance of properly categorizing restoration expenses. The paperwork may seem tedious and unnecessary until they understand the purpose.

One of the problems with emergency account numbers is the potential for abuse. A small stain in a carpet may result in total replacement. A couple of downed trees could be twisted into a complete landscaping overhaul. You have a responsibility to take necessary actions to ensure this doesn't happen. Corporations are under scrutiny because of creative accounting procedures. Insurance fraud is serious business and it's not worth the bad press and possible jail time. To ensure that the emergency account number is not abused, specific instructions concerning its use must be given. Start by providing employees with detailed guidelines about what can and cannot be categorized as a restoration expense. Don't forget to explain the consequences if they elect to abuse the emergency account number.

In the rare event of two different emergencies occurring within a short period of time, you must assign two unique emergency account numbers for each event. The added paperwork may seem a waste of time, but it is a legal requirement. In a two-event scenario you might be responsible for two different deductibles. In an effort to limit financial responsibility, we may attempt to categorize all damage under a single event, but most insurance providers are aware of this possible loophole. This is a concern because most insurance policies contain "misrepresentation and concealment" clauses that void the policy if there is any misrepresentation, fraud, or concealment of material facts during application or future claims. Simply put, if you knowingly misrepresent a claim, your policy provider can cancel your policy and reject all claims without any responsibility. This is not just a threat; numerous cases have been brought to court across the country, and the courts have sided with the insurance provider and not the policyholder.

An interesting twist to this multi-event question has been raised post-September 11. Silverstein Properties, the master leaseholder of the World Trade Center property, is in litigation with its insurance provider, Swiss Re. The debate revolves around Silverstein's belief that the terrorist activities constitute two separate events. Two airplanes were flown into two different buildings at two different times. Eventually both buildings collapsed, but at different times. Based on this belief, Silverstein Properties would be entitled to two separate claims. If the insurance providers have their way, the terrorist attacks would be considered a singular event and, because the policy had a cap of $3.6 billion, they would be responsible for that amount only. Silverstein Properties contends that $3.6 billion would not be sufficient to rebuild both buildings. Property owners and managers along with insurance providers will watch this case with real interest as precedence will be established for future cases.

At some point, you'll need to close the emergency account number down. When you do, provide employees with adequate notice, as some invoices may not have been processed yet. Your organization's accounting department should be prepared to distribute additional information to employees about the emergency account number and its eventual closing.

Catering

In the event of a major emergency, the likelihood of local restaurants being open for business will be minimal. Regarding food and water, emergency team members and employees assigned to assist should be self sufficient for the first 24 hours at a minimum. For ease of use, food provisions might include "meals ready to eat" (MRE). These self-contained meal packages have a long shelf life and do not require heat or refrigeration. Since the employees are responsible for their own provisions during the first 24 hours, the company is able to focus its efforts on the need thereafter.

If the emergency event is major and the recovery effort is likely to last longer than a few days, outside assistance with meal preparation will be necessary. The caterer will need to provide meals for a predetermined amount of people. The vendor should also be responsible for the tents, chairs, utensils, ice, and transportation to the recovery site. It may be necessary to contract with a primary vendor and have an alternate available in the event the primary is unable to service your needs.

Providing Sleeping Accommodations

Everyone needs a good night's sleep, especially when working long hours during the restoration phase. If the emergency was regional in nature, it may be very difficult to find adequate sleeping accommodations for your employees and contractors. Hotels will fill up quick, so early after the emergency, start contacting hotels about availability. If necessary, reserve a block of rooms; it is better to be over-prepared than under-prepared. Because of room availability, it may be necessary to have employees share rooms. If rooms are scarce and your employees are working around the clock, perhaps you could stagger sleeping room usage. For instance, in a room with two double beds, you could comfortably sleep two people. The two people working during the day shift would arrive at the hotel around 7:00 P.M. and leave by 6:00 A.M. the next morning. At around 7:00 A.M. two employees working the night shift would arrive to catch up on showers and sleep. By 6:00 P.M. they would leave the hotel and report back to work. By staggering the use of the hotel room, more people can use a room than under normal circumstances. These accommodations should only be used if additional rooms are not available, and the hotel manager should be made aware of the situation. Logistics such as cleaning, towel replacement, and changing of the linens should be arranged with hotel management.

Some employees may have tents, campers, or other recreational vehicles that they are willing to use during the restoration phase. Will you allow those on site? Is their presence safe? Will they interfere with the ongoing work? Is room available in the building for employees to sleep? Will you be providing cots for workers wanting to take a quick nap? Some organizations have brought in a doublewide trailer and filled it with sleeping cots for workers hoping to catch up on sleep. This may be a good idea, but be sure to get temporary clearance from your local code enforcement body.

Regardless of the provisions you make, you'll want to reiterate company policy on such issues as sleeping room guidelines, shared rooms, per diems, mileage, compensation, and anything else that might come into question. Based on the emergency and the protracted restoration efforts, your organization may find it advantageous to temporarily change some of the guidelines. Publish these temporary changes and distribute to all employees as soon as possible.

Environmental Issues

When a large scale disaster strikes, the damage can be incredible. Buildings can be ripped apart. Rising floodwaters can introduce moisture into walls, resulting in mold and mildew growth. If not taken care of, the growth could become a real indoor air quality issue, causing health problems well after the floodwaters have subsided. Broken or backed up sewer lines create an obvious health issue. Extreme emergency events require much more than a thorough cleaning. The cleanup required after a disaster can be time consuming, expensive, and even dangerous.

If a building experiences even a small level of structural collapse, asbestos contamination could be an issue, especially if the building was built prior to 1980. Think for a moment about possible sources of asbestos in your building. In many old buildings the nine-inch vinyl flooring tiles contain asbestos. The mastic that was used to adhere the tiles to the slab also contained asbestos. Certain types of insulation commonly found in mechanical rooms contains asbestos materials. Ceiling tiles and plaster walls may also contain asbestos. Before becoming overly concerned with the presence of asbestos, understand that asbestos is not considered dangerous until the asbestos fibers are introduced into the surrounding environment. Let's use the example of asbestos floor tiles. Even if your building contains these asbestos tiles today, it is probably not an issue. The asbestos fibers are encapsulated within the tiles. There is no danger until the fibers are disturbed and escape into the surrounding environment. Regarding the mastic, the asbestos contained in it is also encapsulated. No danger exists until the tiles are lifted from the floor and the mastic is disturbed. At this point the asbestos fibers are friable and you now have a very real environment concern.

Now let's think about what may happen to the building during a major disaster. Let's assume an earthquake strikes and we experience a partial building collapse. A large portion of the roof falls in, breaking the asbestos ceiling tiles. The building's foundation was disturbed by the shock waves, causing the asbestos floor tiles to pop up off the slab. Where the floor tiles remained, many were cracked or completely shattered. A few of the interior plaster walls were damaged as well. Based on this building assessment, you have a real problem. Any workers entering this building risk asbestos exposure. If we are not aware of the danger, we may send employees into the building to sweep up the broken tiles. We might instruct a contractor to remove and rebuild the non-load-bearing wall that has been damaged. You have exposed these workers to dangerous asbestos fibers. In our haste to get back to business, we forgot about the environmental concerns of a disaster.

Important Asbestos Definitions

Asbestos Containing Material (ACM)

A term used to describe any material containing more than one percent asbestos.

Presumed Asbestos Containing Material (PACM)

A term used to describe thermal system insulation and surfacing material found in buildings built prior to 1980. Describes materials that may contain asbestos but testing has not been completed to verify its presence.

Surfacing Material

Material that is sprayed, troweled, or otherwise applied to surfaces. These surfaces might include acoustical plaster, fireproofing materials sprayed on structural members, or other materials applied on surfaces for the purpose of fireproofing or acoustical soundproofing.

Thermal System Insulation (TSI)

Asbestos containing material that is applied to boilers, pipes, fittings, tanks, ducts, or other such components to prevent heat loss or gain.

If you have buildings built prior to 1980, and asbestos removal has not been conducted, you should create a written process for safe entry and repairs after an emergency that causes damage to the structure. This process would include establishing a contract with an industrial hygienist who could quickly evaluate a building and provide testing and recommendations. Additionally, employees should be trained to understand the dangers of accessing a damaged building without proper personal protective equipment (PPE) as specified by the industrial hygienist. If employees are allowed to enter the building with specified PPE, OSHA requires employers to provide training and fit testing for the employees. Don't fall for the false belief that a simple dust mask can protect employees from asbestos fibers. These masks are created for nuisance dusts and fall far short of protecting workers from asbestos fibers. Donning inferior personal protective equipment may be the equivalent of fighting a forest fire with a garden hose.

For some organizations, an emergency event could result in a loss of stored hazardous materials. A storage tank could be ruptured and release its contents into the environment. The contents might be toxic gasses, diesel fuel, oil, or anything in between.

> Regarding personal protective equipment, OSHA specifies the required respirators for different asbestos concentrations at CFR 29 1910.1001(g)(3) Table 1. Because of the technical nature of these standards, most employers will not be able to interpret the requirements and supply the appropriate respirator equipment. The testing and specification of personal protective equipment should be preformed by an approved industrial hygienist.

Organizations that have storage tanks should take precautions so the contents will not be easily released. Consider engineering solutions such as containment walls and slabs, alarm systems, and fencing to limit access to these tanks. These precautions should be followed up by procedures to be used in the event of a spill. These corrective procedures might specify the use of absorbent pads, booms, or skimmers. Additionally, contractors specializing in testing, containment, cleanup, abatement, and soil remediation should be identified and included as part of your emergency action plan. By having these contracts in place prior to an emergency event, you ensure quick response to the problem, possibly limiting the impact of the spill.

Assisting Affected Employees

When an emergency has passed and an organization begins its restoration phase, it will often turn to its employees. The best employees have a sense of ownership and they take the rebuilding of the business personally. They can be an excellent source of leadership through the difficult times. They band together and accomplish the unimaginable. But sometimes they get burned out. Sometimes the prolonged stress is too much. Sometimes they don't know how or when to stop and rest. The emotional trauma of a major disaster can cause depression and physical illness. Employers have a responsibility to be aware of the physical and emotional well being of their employees after an emergency event. At a minimum, employers should observe their employees who are involved in the restoration phase. If they notice irritability, withdrawal, or some other behavior out of the ordinary, they must ask the employee to take time off and rest.

If the emergency event is severe, employers have a responsibility to offer crisis counseling to their employees. If the employees do not get well, how can the employee be of value to the company? That may sound inconsiderate, but think about it for a moment. If an employee is physically ill, they stay home from work and recuperate. During their recuperation period, they are not productive. Employers understand that; in fact, they give sick days for just this reason. It stands to reason that when exposed to emotional trauma, we can become emotionally ill as well. We need time to recuperate. We are not productive when in this state. When the trauma is great, crisis counseling may be necessary. An organization that is truly concerned about business continuity and the well-being of their employees will ensure that the employees are physically and emotionally well.

If an emergency is regional in nature, the damage will not be limited to your organization's business or property. Neighboring business will also be affected. The homes of employees may be damaged. Employees or their family members could be injured or even killed as a result of the emergency event. Family concerns should always take precedence over business. Realizing this, a caring organization would not expect employees personally affected by the emergency to be available for the restoration phase. They will have their hands full dealing with their own loss. In these cases, it makes sense to bring employees in from different locations that were not affected as deeply by the emergency.

In the worst disasters, some employees may be rendered homeless. As an organization, you'll want to assist, but realistically what can you do? This is not an issue that can be answered for you. It is unlikely that your organization will be able to purchase new homes for employees and that would not be expected. However, can your organization assist employees with temporary housing? Perhaps your organization can guide employees to federal or state assistance. The assistance is out there, but many people don't know how to access it. Can your company organize federal assistance workshops for displaced employees? Can your organization help employees with the mounds of paperwork required by insurance providers? Again, the level of service you provide will be a corporate decision, but remember, business can't get back to normal until the employees get back to normal. Employees are an organization's greatest investment and should be treated as such.

Don't Forget about Safety

Most employers are required to adhere to OSHA regulations, including specific recording and reporting responsibilities. The possibility of a workplace injury is increased while recovering from an emergency event, and for this reason particular attention must be given to employee and contractor safety.

In the event of an accident that results in any injury, even a minor one, a post-accident report should be completed. The purpose of the post-accident report is to identify all injuries, even those that do not trigger an OSHA recordable. Ultimately, the post-accident report form will serve as tool for trending analysis. If employees keep getting injured while moving sandbags, a proactive manager would look for ways to eliminate or reduce the risk involved. If the injury falls into OSHA's recording regulations, the injury and associated information must be transferred to the appropriate OSHA record keeping.

The post-accident report typically includes the following information:

- What happened?
- What activity were the employees engaged in?
- When did it happen (date and time)?
- Who witnessed the event?
- What were the weather conditions?
- Was the accident due to an unsafe condition or unsafe act?
- What preventative steps will be taken?

Although not required by OSHA, the post-accident report shows your organization's desire to proactively reduce workplace injuries. OSHA has created three specific forms that must be completed by employers when specific injuries or illnesses occur. Some industries and organizations employing ten people or fewer are exempt for OSHA's record keeping requirements unless selected by the Bureau of Labor Statistics to participate in an annual survey. The required OSHA forms include

> **Don't Forget:** OSHA Form 300A (Summary of Work-related Injuries And Illnesses) must be completed and be displayed next to the OSHA safety poster (Form 3165 or state equivalent) from February 1 through April 30 each year.

- **OSHA Form 300** (Log of Work-related Injuries and Illnesses)
  OSHA's Form 300 is used to record information on all illnesses and injuries that require medical attention beyond basic first aid. The OSHA Form 300 is used to record information such as the employee's name, job title, date of injury, description of injury, and the location where the event occurred. The OSHA Form 300 replaces the old OSHA 200 form in 2002 and must be kept for five years.
- **OSHA Form 301** (Injury and Illness Incident Report)
  The OSHA Form 301 is the individual record of each work-related injury or illness recorded on the 300 Form. It includes additional data about how the injury or illness occurred. The 301 Form replaces the former OSHA 101 Form.
- **OSHA Form 300A** (Summary of Work-related Injuries and Illnesses)
  The OSHA 300A replaces the summary portion of the former OSHA 200 Log and is now a separate form, updated to make it easier to

calculate incidence rates. It must be signed by a company executive. Even if your organization does not have any recordable injuries, the OSHA Form 300A must be completed. Fill in all the blanks and add zeros in spaces that ask for total injuries and illnesses.

Regardless of the severity, all occupational illnesses must be recorded on the OSHA Form 300. An illness is defined as an abnormal condition resulting from something other than an occupational injury, caused by an exposure to environmental factor at the workplace. These factors include illness caused from inhalation, ingestion, and direct contact. Examples of illnesses can be categorized into four groups:

Skin disorders:
- Dermatitis
- Rash caused by poisonous plants
- Inflammation of the skin

Respiratory conditions:
- Silicosis
- Asbestosis
- Tuberculosis

Poisoning:
- Lead
- Mercury
- Carbon monoxide
- Insecticide

Other:
- Heatstroke/Sunstroke/Heat exhaustion
- Frostbite
- Anthrax
- AIDS/HIV
- Hepatitis B
- Mold

Reporting requirement. Within eight (8) hours after the death of any employee from a work-related incident or the in-patient hospitalization of three or more employees as a result of a work-related incident, you must orally report the fatality/multiple hospitalization by telephone or in person to the Area Office of the Occupational Safety and Health Administration (OSHA), U.S. Department of Labor, that is nearest to the site of the incident. You may also use the OSHA toll-free central telephone number: 800-321-OSHA (1-800-321-6742).

Most injuries and illnesses are simply recorded on the appropriate OSHA logs and made available for display during the months of February, March, and April. You are not required to contact OSHA by telephone unless a death or the in-patient hospitalization of three or more employees as a result of a work-related incident occurs.

When you contact OSHA to report an employee death or in-patient hospitalization of three or more employees as a result of a work-related incident, OSHA will ask the following questions:

- The establishment name
- The location of the incident
- The time of the incident
- The number of fatalities or hospitalized employees
- The names of any injured employees
- The name and phone number of a contact person
- A brief description of the incident

When evaluating an injury for OSHA record keeping purposes, always ask youself, "Does this injury require medical treatment beyond basic first aid?" If the answer is yes, you must record the injury in the OSHA Form 300. For those areas considered questionable, the following list of medical treatments may be helpful:

- Treatment of infection
- Application of antiseptics during second or subsequent visit to medical personnel
- Treatment of second or third degree burn(s)
- Application of sutures (stitches)
- Application of butterfly adhesive dressing(s) or steri strip(s) in lieu of sutures
- Removal of foreign bodies embedded in eye
- Removal of foreign bodies from wound; if procedure is complicated because of depth of embedment, size, or location
- Use of prescription medications (except a single dose administered on first visit for minor injury or discomfort) Use of hot or cold soaking therapy during second or subsequent visit to medical personnel
- Application of hot or cold compress(es) during second or subsequent visit to medical personnel
- Cutting away dead skin (surgical debridement)
- Application of heat therapy during second or subsequent visit to medical personnel
- Use of whirlpool bath therapy during second or subsequent visit to medical personnel
- Positive X-ray diagnosis (fractures, broken bones, etc.)
- Admission to a hospital or equivalent medical facility for treatment.

Additionally, these incidents are always recordable:

- Loss of consciousness
- Restriction of work or motion
- Transfer to another job
- Death

If one or more of these events have occurred, you must record the injury. On the other hand, administration of simple first aid does not trigger an OSHA recordable. Events that would be considered first aid treatment would include

- Application of antiseptics during first visit to medical personnel
- Treatment of first degree burn(s)
- Application of bandage(s) during any visit to medical personnel
- Use of elastic bandage(s) during first visit to medical personnel
- Removal of foreign bodies not embedded in eye if only irrigation is required
- Removal of foreign bodies from wound; if procedure is uncomplicated and is, for example, by tweezers or other simple technique
- Use of nonprescription medications and administration of single dose of prescription medication on first visit for minor injury or discomfort
- Soaking therapy on initial visit to medical personnel or removal of bandages by soaking
- Application of hot or cold compress(es) during first visit to medical personnel
- Application of ointments to abrasions to prevent drying or cracking
- Application of heat therapy during first visit to medical personnel
- Use of whirlpool bath therapy during first visit to medical personnel
- Negative X-ray diagnosis
- Observation of injury during visit to medical personnel

Keep in mind, however, that the loss of consciousness, restriction of work or motion, or transfer to another job, even if the employee did not receive medical attention does trigger an OSHA recordable.

Now Is Not the Time to Let Your Guard Down: Security

Immediately following an emergency event, most people will be totally consumed with finding victims, mending their own wounds, making repairs to their buildings, or restoring business continuity. When we become so involved in these important matters, we sometimes miss the obvious around us. We may not be as perceptive as usual, overlooking the person with bad intentions. In life, and especially after an emergency, some people will prey on the misfortunes of others. We assume that after an emergency everybody wants to pitch in and help, but this is not always true. Even those who might not normally be inclined to dishonesty, in a moment of weakness and under the guise of providing for their family, may view your misfortune as an opportunity to get something for nothing.

In one example, a delivery of fifteen portable generators was sent to a power plant immediately following a hurricane. The generators were left in a semi-secure area, but within four hours they had disappeared. The employees of the power company were unable to watch the generators and the security guards who were called to duty were delayed in reaching the plant. In another example, immediately following the final game of an NBA championship series, fans of the winning team began to celebrate. Most of the celebration was harmless fun, but a few of the revelers had bad intentions. These people smashed shop windows and began looting stores. Liquor, furniture, and electronics were taken. Because of situations like these, you cannot take security for granted.

Because we never know when an emergency will occur, security contracts must be in place continuously. Writing a good contract will require you to ask yourself and the security provider a number of questions:

- What background checks will be done on the guards?
- How quickly will the guards respond?
- Can guards be brought in from another area not affected by the emergency?
- Will you need security around the clock or just during evening hours?
- Do you require the guards to be armed?
- Is there any other specialized training that is needed?
- To whom will the guards report?
- Where will they be stationed?

One of the key issues will be availability of guards immediately following the emergency. If the emergency event was regional in nature, an earthquake for instance, the usual guards may be unable to respond because phone service is out or they are assisting their own families. Can the security company you are contracting obtain additional guards from a nearby area not affected? If so, how long will it take them to arrive? It's often advantageous to have an alternate security company available in the event the primary company is unable to respond to your needs.

Monitoring Restoration Progress

Because of the psychological trauma a major emergency event causes, the restoration progress can be delayed. Often the work seems overwhelming. There appears to be no end in site. If you speak with people who have witnessed great destruction they will tell you that the damage is so great you sometimes don't even know where to start. After Hurricane Andrew struck South Florida, people were wandering around in a daze, unable to take care of (or unaware of) their injuries. People were without homes. They spent their day scavenging for food. They spent their evening sleeping in piles of rubble.

When this level of loss is experienced, work and the restoration phase is secondary. You may find that members of your emergency management team, restoration contractors, and employees have experienced great personal loss as a result of the emergency. They may not be able to focus on the large restoration task at hand. They may be easily distracted or they may be concerned with the welfare of their families. For this reason, be very selective about who you choose to monitor your restoration progress. You'll need to have a leader on board who can remove himself or herself from the emotional trauma of the event. This person may be an employee from a non-affected area or it may be an outside contractor. Regardless, make sure the person is emotionally able to handle the task.

Like any large project, the restoration phase must be kept on track. If left unattended, valuable time will lapse without significant progress being made. This is especially critical in cases in which business continuity is at stake. If your organization is unable to conduct business, and for every day it is unable to do so, the company loses $250,000, the impact is significant. After the emergency, building assessments must be performed. Based on your findings, you may decide to bring in an architect or structural engineer. At some point a scope of work should be decided upon and timelines should be developed. A project manager is critical if the required restoration is substantial. In fact, you may elect to bring in a project manager who specializes in recovery from disaster events. Such people have unique experiences dealing with restoration projects and can be instrumental in locating federal assistance that may be available to your organization.

Because of the stress of the restoration phase, people don't always act in ways that are normal. The slightest problem could blow up to be a major event. Disagreements will occur on any jobsite and usually the problems are worked out without a major conflict. But think for a moment about the restoration phase. A major disaster has just occurred. People are stressed. They haven't eaten or slept properly in days, perhaps weeks. They are working long hours, sometimes twenty hours a day, for seven days each week. They have been displaced and they haven't seen their families for weeks. Would you be a little edgy? Could you, in a moment of weakness, snap at somebody if you felt they weren't working as hard as you? We're all human and we should be aware of the negative emotional effects following a major disaster.

Safety is another concern after an emergency event. Many people have been unnecessarily injured or even killed during the restoration phase because of a lack of safety measures. In our zeal to get back to a sense of normalcy, we take shortcuts and practice unsafe work habits. Our property may not be structurally sound, but we enter it. Floodwater may create a breeding ground for mold and other indoor air quality issues, but we breath the air without a respirator. Employees are eager to help out, but are not properly trained or clothed. Broken glass is strewn about the floor, yet we walk on it without proper work shoes. Do not minimize the risk associated with the recovery efforts. Insist that all workers—employees or contractors—don the required safety gear. Make sure that all workers are properly hydrated, fed, and protected from the elements. Conduct regular safety reminder meetings and stop those workers who might be working in an unsafe manner. It would be sad if an employee survived the emergency (over which they had no control) yet lost his or her life during the recovery stage (when he or she did have control over the situation).

Since we mentioned the negative aspects of emergency restoration, we should also give equal time to the positives, and there are many. The human spirit can accomplish wonderful things. When morale seems to be at its lowest, when the day is darkest, heroes are born. Normal everyday people stand up and do miraculous things. Take for example the World Trade Center attacks on September 11. Nearly 3,000 people lost their lives, shocking not just the country but also the entire world. Immediately, trained rescue personnel from around the world joined in to search for victims. Eventually, search and rescue became search and recovery. The task of removing the rubble seemed impossible. Experts estimated the job would take over a year to complete. In just eight and one-half months, nearly two million tons of debris were moved to a Staten Island landfill. Over three million hours of labor were spent to conduct the clean up of the WTC site. The world watched with great interest as the community rose up and made unprecedented progress. The city could have thrown up its arms in despair. City leaders could have argued about the best way to perform the recovery. The world was watching, and what they witnessed was beyond belief. The compassion, industriousness, ingenuity, and resolve displayed by those working in the recovery phase will never be forgotten. Did the terrorists win? I don't think so.

Keeping the Lines of Communication Open

The flow of factual and comforting information is very important, especially following an emergency event. People are confused, even scared. Rumors run rampant. The emergency planner will play an important role in disseminating reliable information. Whether you are communicating with employees or the media, concise and truthful information must be shared.

A number of forums for communications exist:
- Radio or television advertising
- Written press releases
- Press conferences
- Interviews
- Site tours
- Telephone hotlines
- Corporate memos
- Website updates
- Panel discussions
- Community meetings

These forums should be selected based on the severity of the emergency, the purpose, and the target audience.

Communication with the Media

The media will seek statements from your organization. Concerning the media, the worst action you could take is inaction. If you fail to return calls or respond to questions, you run the risk of the media consorting with other sources such as competitors, employees, or neighbors. These sources may be biased or uninformed, and false information could be delivered. So rule number one is: Don't avoid the media. Rule number two is: Don't say more than you should. No matter how articulate you might be, no matter how comfortable you might be in front of an audience, you absolutely must plan and practice the press conference. Practice in front of a "friendly" audience first. An audience of trusted employees in your organization would think of all the hard questions that could possibly be asked. Discuss what should and should not be said. Draft out responses to the questions you know will be asked. For instance, you can count on the following questions being asked:

- What happened?
- When did it happen?
- Were there any injuries?
- How many?
- Were there any deaths?
- How many?
- Why did the accident happen?
- Will neighboring areas be affected?
- When will the situation be back to normal?
- What assurances can you give us that this will not happen again?

It is important to remember that reporters have a job to do. Their timing might not be the best. They may come across as pushy or trouble-makers. You can turn the negative press conference into a positive event. Instead of viewing the conference as intrusive and a waste of time, think of it in terms of free advertising. You can advertise your organization in a positive light, one that is proactive and concerned with the safety of its employees and the community. Imagine if your organization purchased a thirty second advertisement on radio or television. It would be expensive, no doubt about it. During a crisis, you can take advantage of the free advertising afforded to you because of the news coverage, but make sure it is good advertising! Following an emergency, many organizations were recognized and awarded for their resolve in the face of adverse conditions. They took advantage of media available to them, and so can you.

Will you allow the media onsite? That depends on the safety of the site and your personal comfort level. Realize, though, that keeping reporters off the site for reasons other than safety will raise a red flag. If you label the site unsafe, you'll likely be met with doubters. It may be advantageous to have a fire safety spokesperson or an engineer available to back up your claim. If the site is truly unsafe, make sure the affected areas are taped off and warning signage is visible. You may also elect to supply photographs or videos of the inaccessible areas. Samples of the damage will also provide the non-technical reporter with a clear understanding of the problems and the efforts to repair the facility. If you elect to bring the media to your site, arrange for a media briefing area. Do not allow media representatives to walk unescorted through your site.

Because an extended emergency event may require multiple communications with the media, you must keep detailed records of information released. When communicating verbally with the news media, also provide them with a written press release when possible. In the spirit of speaking to your target audience, avoid the use of technical terminology. Use "plain" English. If you find it necessary to provide technical information, complement that information with a written press release that will be handed out at the press conference. To clarify technical points, consider using charts, graphs, pictures, and videos to get your point across.

Because of the delicate nature of corporate communications, a few words of warning are in order. First, do not speculate about the incident or its causes. If you don't know the answer to a question, promise to keep the media informed as more information becomes available. Do not attempt to place blame or cover up facts. You must develop a clear chain of command regarding corporate communications. This can be accomplished by limiting the individuals with responsibility as company spokesperson. Typically one person is assigned spokesperson duties and one or two others are assigned as alternates. Finally, remind employees to direct all questions from the public to the company media relations department. If such a department does not exist, create one on a temporary basis to accommodate the emergency event.

Communication with Employees

Fear, anxiety, and, in some organizations, distrust are natural byproducts of an emergency event that requires extended recovery. Employers have a responsibility to communicate with employees when emergency-related events occur. Employees want to know when it is safe to return to work. They may need additional assistance if they were directly affected by the event. In the event of an accident, employees want to know what has been done to ensure that the accident is not repeated. Quick and helpful information is an effective way for organizations to show that management cares for the well-being of the employees.

After an emergency event, communication should not be limited to verbal announcements. Written communication should also be provided to the employees. Although not actually a press release, the formal communication is an invaluable tool for the employee/employer relationship. Without communication, employees will turn to rumors and innuendo. Issues such as work locations, return-to-work dates, and compensation during restoration should be addressed. Reminders on how to handle media inquiries and rumors should also be covered. Keep the communication positive and always thank the employees for their dedication.

Communication with Families

When emergency events occur and the ramifications are protracted, the families of employees can be affected. Anxiety and fear can paralyze those who do not know the condition or whereabouts of their family members. If the emergency event causes extended financial hardship on the organization, employees and their families may be concerned about job security. A delay in delivery of a paycheck could prove disastrous for many families. If layoffs are necessary, employees and their families will be interested in severance packages, health insurance continuation, and job placement assistance. These concerns must be addressed as soon as possible.

Communication with Insurance Providers

Although insurance providers will send a claims agent to review damages to your property, you may consider a proactive approach to communication to be wise. This is especially important when an emergency is regional in nature and the insurance provider is overburdened with claims. A formal communication describing the event and how you were affected is suggested. The correspondence should describe the damage incurred and identify the employer's contact information.

Communication with Government Agencies

An often overlooked target audience is government agencies. These agencies might include FEMA, Red Cross, police, National Guard, FBI, fire rescue, and even OSHA. Formal written communication may be recommended based on the nature and impact of the emergency event. For instance, an on-site oil spill would warrant formal communication with the Environmental Protection Agency. They will be interested in what happened, when it occurred, what is being done to address the spill, and what measures are being taken to ensure the incident is not repeated. In scenarios such as this, it is better (and often required) to contact them before they contact you.

Communication with Customers

Contact with customers is often overlooked after an emergency event. If the emergency event is great and critical operations are impacted, customers may be impacted as well. Your customers will have concerns about your ability to provide services or products in a timely fashion. They may also be concerned with your organization's ability to pay for services rendered. Remember too, your customer may also be the shareholders in a corporation. They have some very real concerns about their investment that should be addressed. Because of these legitimate concerns, you must be proactive in communicating with them in a formal manner.

Sample Press Releases

The first example is a typical press release format. Simply fill in the appropriate information for the particular event.

For Immediate Release

From: [Your company name]

Address: [Your address here]

Date: [Today's date here]

Contact: [Name and phone number of company representative]

[Sample text]

On September 22 at approximately 2:30 a.m., a Category 3 hurricane swept across the southern Louisiana area including New Orleans.

Allied Industries has three properties in the affected areas, two of which were damaged. The affected properties are located at 123 South Street and 87 Canal Street.

The extent of the damage appears to be minimal. The properties are currently being assessed and further information will be made available as it arrives.

Restoration of the affected buildings will commence immediately after the assessments are completed.

There were no injuries and all employees have been accounted for.

END OF RELEASE

This next example is an internal memo from the president of a corporation. The target audience in this scenario is the company's employees; therefore, it warrants a different style. New information and reminders of company policy are addressed in this example.

Dear Fellow Employee,

Our business was built on hard work and dedication from our employees. That need for hard work and dedication has never been as evident as it is today. Your efforts are greatly appreciated and I have no doubt we will persevere. We are faced with an excellent opportunity to show our resolve and character, and I invite you to show your best.

When faced with unusual events such as what we have just experienced, a few reminders are in order. We will keep our employees updated via e-mail and fax service. For those unable to access these mediums, we have established an emergency hotline (800-123-1234) with recorded updates and frequently asked questions. We realize that rumors and innuendo will surface. We ask that you do not perpetuate such comments. All official communication will be provided through the methods listed above.

For many of our employees, the media will be observing your work. Again, you have an opportunity to present the company in a positive way. Unfortunately, some media representatives may attempt to sensationalize disasters and how they are dealt with. If asked about your restoration efforts, you may briefly explain what you are doing, stressing the fact that your efforts are part of a larger restoration effort by all of the employees of the company.

We ask that you refer questions involving detailed aspects of your work to our corporate communication department at (800-555-1212). Finally, if you don't know an answer to a reporter's question, don't speculate. It's ok to say, "I don't know." If you find yourself in an uncomfortable situation, thank the reporters for their time, and kindly remind them that you need to get back to work.

I am very proud of the efforts of our employees and I know we can all count on each other to make a complete recovery a reality. Remember, we have an opportunity to make a difference and your efforts are greatly appreciated.

President

Acme Industries

Summary

After an emergency event has occurred, a successful restoration phase can be the difference between business continuation and bankruptcy. The employer will be faced with many difficult decisions, and how quickly these decisions are made becomes critical to survival. Confident and quick communication with the media, suppliers, clients, and other groups must not be overlooked.

During the restoration phase employee comfort and safety must not be minimized. Employees will be concerned about their jobs. Many employees will be affected by the emergency at home, and the stress of the events can be overwhelming. Employees involved in the restoration phase will likely be working long hours, and in their zeal for recovery, will not always take proper precautions or get proper sleep. These are some of the issues that must be addresses during the restoration phase.

8

Guidelines for Emergency Mitigation

Design Guidelines

As security and emergency concerns grow in importance, commercial real estate must adjust accordingly. What worked 50 years ago won't work today. For that matter, what worked ten or even five years ago may now be obsolete. The bad news is some emergency events will never be eliminated. The good news is the negative effects of just about every emergency can be reduced. The reduction of these events or effects starts with proper planning. Most of this book discusses planning from a what-do-we-do-if-"X"-event-occurs frame of reference; however, planning in the design phase is an equally important mitigation factor. Proper design of the building structure and surrounding site is critical, especially in those buildings that are open to the public.

When discussing building and site security, it makes sense to separate the space into five distinct areas, working from the inside outward:

- Interior—Limited access
- Interior—Public access
- Point of Entry
- Exterior—Building perimeter
- Exterior—Parking and other off-site areas

Interior—Limited Access

The interior limited access areas include spaces such as storage rooms, telecommunications areas, computer server rooms, and mechanical rooms. These areas are off limits to most people, including visitors and employees. Access can be controlled through the use of keys, proximity cards, mag-lock, card swipe system, or any other feasible manner. Start by performing an analysis of your building's space and determine what spaces need to be off limits due to security and safety concerns.

Interior—Public Access

In most buildings, especially those that offer public access, the amount of limited access interior space will be minimal. Even if this is true in your building, care must be taken to secure areas open to the public. For instance, can visitors to the building gain access to the roof? Are they able to access HVAC systems including intakes and discharges? Could a visitor with bad intentions force his way into limited access areas? By performing an assessment of your building's public spaces, you'll be able to identify areas of concern.

Point of Entry

Spending large amounts of money and time securing the inside of a building without addressing points of entry could be akin to changing the oil in an automobile after the engine has blown. It might be too little too late. What will you do to ensure that bad people and their tools of destruction don't gain access into your building in the first place? The obvious starting point is to have some way to police the entrances. This can be accomplished by employing security guards, card access, employee escorts, or a number of different solutions. For after business hours, an intrusion detection system should be installed. However, the danger lies in assessing your building's point of entry security from a doorway access point of view only. Think about other less obvious points of entry such as windows, loading docks, fresh air intakes, and other such penetrations.

Exterior—Building Perimeter

The immediate exterior building perimeter can also be an area of weakness. Planters, overhangs, and other building features could be the perfect place to hide an explosive device. When performing a security assessment take these areas into consideration. Is the immediate perimeter lighting adequate? Can visitors with bad intentions gain access to or sabotage utilities or the emergency generator? Could a toxic substance be introduced into your building via a fresh air intake? Does the perimeter landscaping create a possible hiding place for criminals or explosives?

Exterior—Parking and Other Off-Site Areas

A person with bad intentions may not even need access inside of your building to do damage. In the case of the Alfred P. Murrah federal building in Oklahoma City, the terrorist was able to bring down most of the building and kill 168 people without ever entering the facility. By parking a van full of explosives near the building and detonating them from a distance, incredible damage was inflicted. Studies have been performed that suggest standoff distances of 100 feet or more will lessen the impact of a bomb. Since the Oklahoma City bombing, we realize the importance of not allowing unattended vehicles to park in front of buildings. Keeping this 100 foot standoff distance in mind, can you keep distance between your parking lot and the building? If you do, you may wish to make special arrangements for those with disabilities because the distance may create hardships. Parking garages create a dilemma. Since they are a building structure that could fail, an explosion could create tremendous damage. Consider limiting access to your parking garages and installing closed-circuit television to monitor movement. If you employ underground parking, it may be wise to limit it to registered vehicles, motorcycles, and perhaps small cars. Better yet, it may be prudent to eliminate that parking area altogether if feasible.

Site design considerations for bomb resistance might include:
- Bollards/planters
- Elevation of building level
- Curbs
- Security lighting and cameras
- Gates

The purpose of protective bollards, planters, or even large pieces of exterior artwork is to keep a vehicle from gaining close access to a building and detonating explosives. Although bollards and retaining walls are effective, they can be aesthetically unattractive if not designed properly. The use of natural barriers such as trees with thick trunks can prove to be an efficient barrier that is aesthetically pleasing as well.

When reviewing your building and the desired level of security, always remember ascetics and the role it plays in employee satisfaction and desirability of public interaction. You could build an intimidating fortress without windows which affords a great measure of security; however, is that really desirable? Would employees enjoy working there? Might their productivity be reduced in such an environment? Would the public be too intimidated to visit and do business there? On the other hand, you could design a high-tech, functionally proper, aesthetically pleasing masterpiece. But at what cost? Anything can be accomplished if the price is right, and when it comes to a totally secure and safe structure, the price will likely be too high. The trick is to find an acceptable balance of risk and cost. Performing a risk analysis would be a prudent starting point.

> To learn more about the science of CPTED, contact Atlas Safety & Security Design, Inc.
>
> The Atlas website features some very interesting information on CPTED and how it can be integrated into both remodels and new buildings alike.
>
> Contact: Randall Atlas Ph.D., AIA, CPP
>
> Atlas Safety & Security Design, Inc.
>
> 770 Palm Bay Lane
>
> Miami, Florida, USA 33138-5751
>
> Telephone: (305) 756-5027 or (800) 749-6029
>
> Fax: (305) 754-1658
>
> <www.cpted-security.com>

Consulting with a design specialist who understands security and safety would also make sense. The science of designing viable buildings that are safe and secure is known as "crime prevention through environmental design" (CPTED). Designers and consultants specializing in CPTED bring an added dimension to your building design. Their intimate knowledge of security and safety is integrated with an architecturally attractive and functional building. These consultants will identify opportunities based on your need and budget and typically are well worth the price.

Windows and Doors: The Weakest Link

The windows and doors of a building often present the path of least resistance during an emergency. Whether the event be terror related or weather related, windows in particular are usually the weakest link. Next to total building collapse, more people are injured from flying glass during an explosion in a building. In fact, most people who are injured by flying glass are within fifteen feet of the windows when the blast occurred. A recent example of this phenomenon can be traced to the bombing of an embassy building in Africa. The first explosions were produced by grenades detonated outside of the embassy. The grenades did very little damage but startled the occupants of the embassy. As the occupants of the building began to walk toward the windows to investigate the loud noise, a much larger bomb planted in an automobile was detonated. The tremendous explosion shattered glass and turned the shards into razor-sharp missiles, killing 220 people and wounding nearly 5,000.

When examining the problems with windows, glass is just one of the issues. Many windows are not installed in a way that offers true blast protection. Often during an explosion or storm, the entire window becomes dislodged from its wall opening. A number of safety features are available that can mitigate the damage caused by windows and glass. In some areas, such as south Florida, window shutters are now required to protect people and buildings from the effects of storms. Protective window film applied over existing windows can also provide some protection from pressure experienced during a storm or explosion. The protection, however, is limited. The real purpose of window film is to keep glass from shattering and dangerous shards from becoming airborne. Protective film does resist impact. The glass will break; it just won't disperse as readily as it would if unprotected. If you looking for protection from shards and impact resistance, ballistic grade glass would be an option worth exploring. The cost may be high, but the protection provided, especially in a high-risk building, may well be worth the investment. When installing blast-resistant glass, it is important to understand construction methods for installing the windows. If installed improperly, the window may absorb an impact and the glass would remain unbroken, but the entire window frame would become dislodged from its opening. Installation of blast resistant glass should include extra reinforcing around the window opening. Consult with the manufacturer and structural engineers when installing blast resistant glass.

The General Services Administration (GSA) is the agency that provides and manages real estate for the federal government. Because of the likelihood of a federal building becoming a target for a terrorist, the GSA has created specific standards for glazing installed in federal

> The Pentagon is a good example of a properly constructed, high-risk building. Because of the high profile of the Pentagon, the windows need to be blast resistant. Each window unit weighs about 1,600 pounds and the glass is 1.5 inches thick. The windows are bolted to massive structural tubes that run the entire height of the building. At the bottom of each window, extra reinforcement is provided by a steel plate that is attached to the window and the steel tubes. This would likely be considered overkill for most buildings; however, the Pentagon serves as a model for proper blast-resistant construction. The window components including the glass, window frame, and its installation are constructed in a manner that ensures proper blast resistance.

buildings. Level 1 is the highest category and specifies that "glazing does not break." This would necessitate the installation of ballistic glass. The lowest protection category is Level 5 and is defined as "windows fail catastrophically." When considering the purchase of blast-resistant windows or protective film, speak to the window manufacturer about its ability to meet GSA level specifications.

Protective Bollards, Planters, and Green Space

As experts study the destruction caused by explosions, they have unanimously agreed that "standoff distance" is the best solution for decreasing the impact of the explosion. Simply put, the more green space between the building and the bomb, the less of an effect the explosive will have on the structure. In a practical sense, standoff distances can be increased by installing protective bollards or planters around the building. The bollards are reinforced concrete posts, usually cylindrical, ranging from four inches in diameter to 12 inches diameter or greater. These bollards are very successful in keeping vehicular traffic from entering a restricted area. On the negative side, bollards tend to look unattractive. They look as if they were placed to keep people out. Urban designers and architects agree that protective barriers should blend into the environment. They should not be readily noticeable to the average person. There are barriers cleverly disguised as decorative walls, planters, benches, or artwork. Landscaping and fencing can be used to keep vehicles at bay. Forcing points of entry into a building through a long corridor minimizes the affects of an explosion as well.

> Permanent storm shutters offer the best protection from the high winds of a hurricane. If you live in an area that is susceptible to hurricanes, the investment in storm shutters is a good one. If you do not have permanent storm shutters, the next best option is to construct window coverings using 5/8 inch marine grade plywood. Pre-drill holes every 18 inches for screws and pre-install wall anchors. Don't forget to mark each piece of plywood so you'll know where it belongs on the building.

Virtual Reality Software and Related Technology

Scientists and engineers have realized the importance of recreating events with a goal of improving processes or finding out what went wrong. This study of events can be applied to building failure due to an emergency event. In particular, building simulation and the creation of virtual environments have assisted engineers in determining why a building structure failed. In true proactive fashion, engineers have used this data to better design buildings that can absorb emergency events including fire, earthquakes, terrorism, and other such events.

Typically, computer-aided design (CAD) software is used to create a three dimensional model of a building. This virtual building would then be placed in a digital environment that mirrors the subject property. At this point, programmers can subject the building to a host of virtual events while recording the outcomes of each. Modifications can be made to the 3-D building including ballistic glass, increased standoff distances, or other design features, and the effectiveness of each modification can be measured. This process allows the engineer to select the appropriate modifications for the safety level desired.

> Because vehicles can be used to store and transport powerful bombs, you must take proactive steps to protect your building from these dangers. Begin by not permitting unoccupied vehicles to park in front of your building. Enforcing no parking zones is also suggested. In some scenarios parking permits are issued for employees and visitors must park in a special area that is located at a sufficient standoff distance from the building. It is imperative to contact your local police department if you find unidentified vehicles in your parking lot, and have the vehicle owner identified and the vehicle towed, if necessary.

Although the availability of such software and expertise is currently limited to certain high-risk structures, the value of these studies is evident and growing in popularity. As technology evolves and expertise is developed, the cost of these studies will become more tolerable as well.

Post-Failure Analysis

Immediately after a building collapse, specialized scientist and engineers will comb the premises to collect as much data as they can. To the untrained eye, the pile of rubble where the building once stood seems to contain very few clues of any value. To the highly trained eye, the site is a treasure trove of information. What caused the building failure? Where did the problem start? Could it have been prevented? Could better building materials have averted this disaster? Could better building techniques helped? These are the types of questions that can be answered by performing a post-failure analysis. The collapse of the World Trade Center Towers necessitated a post-failure analysis at an unprecedented level. Well over a year after the World Trade Center collapse, engineers were still sifting through the rubble, searching for any clues that may shed more light on the event. No doubt, we will learn from this disaster and our buildings, particularly high-rise buildings, will be better protected in the future.

Building Codes

Development and implementation of improved building codes has been instrumental in mitigating the impacts of various emergencies. Codes requiring tie down straps on rafters, fire sprinkler systems, storm shutters, or other proactive steps have saved lives and property. As we continue to better understand our exposure and we improve our building techniques, we can expect to minimize the effects of many emergencies.

Because codes are not uniform across the nation, this book will not attempt to examine the individual codes and provide endorsements. Instead, become familiar with the local codes that bind your building and renovation projects. Many of the existing structures across the country were built to inferior standards and therefore present a greater risk if faced with an emergency. When an emergency event, particularly a weather-related emergency, strikes a community, architects and engineers study the buildings that fared well and the buildings that did not fare so well. Time and time again, the older buildings built prior to the sticker codes comprised the bulk of the damaged structures. Because of this fact, when considering the purchase or lease of a building, part of your due diligence auditing should be based on the construction techniques of that building.

One code that is quickly gaining acceptance across the country is based on the NFPA 1, Fire Prevention Code® (2000 Edition) or the NFPA 101, Life Safety Code® (2000 Edition). These resources contain specific requirements for fire drill frequencies in a number of different kinds of buildings. Many local jurisdictions will refer to the NFPA standards and adopt them "by reference." Even those jurisdictions that do not directly refer specifically to the NFPA requirements are using them as a model for their own codes. For these reasons, its worth getting a copy of the NFPA code book and becoming familiar with it.

Building Performance Assessment Team

The Building Performance Assessment Team (BPAT) program is operated by a consulting engineering firm under contract to FEMA. The purpose of the BPAT program is to study building performance in response to all hazards and emergencies. The goal of the program is to better understand what makes a building survive emergencies and how the negative effects of emergencies can be mitigated.

After an emergency event, a Building Performance Assessment Team is dispatched to the emergency location to inspect the affected buildings. The assessment team will conduct engineering analysis to determine the cause of failure, or the reasons for successful structural integrity. At the end of the team's analysis, they will recommend actions that could reduce future damages and protect lives during similar hazards. The positive results of the BPAT program include improved disaster-resistant building codes, design methods, and material selection.

The Building Performance Assessment Team consists of FEMA representatives, state and local officials, and public and private sector technical experts. These technical experts include persons from the fields of

- Structural and Civil Engineering
- Geotechnical Engineering
- Environmental Engineering
- Coastal Engineering
- Retrofitting
- Shoreline and Coastal Erosion
- Flood-Resistant Design and Construction
- Floodplain Management
- Wind-Resistant Design and Construction
- Historic Preservation
- Earthquake-Resistant Design and Construction
- Water Resources
- Hurricane-Resistant Design and Construction
- Building Code Analysis and Evaluation
- Forensic Engineering
- Architecture

For those interested in learning more about FEMA's BPAT program, including how to volunteer and be included in the BPAT expert database, contact

Program Assessment and Outreach Division
Federal Insurance and Mitigation Administration, Room 411
Federal Emergency Management Agency
500 C St., S.W.
Washington, DC 20472
Phone: 202-646-3935
Fax: 202-646-257
<www.fema.gov>

Since the early 1990s, FEMA has deployed BPATs in response to a wide variety of disasters including hurricanes, floods, and the bombing of the Murrah Federal Building in Oklahoma City. To appreciate the thoroughness of the BPAT's work, you need look no farther than the World Trade Center Disaster on 9/11. The Building Performance Assessment Team conducted field observations at the WTC site and steel salvage yards. They meticulously removed and tested samples of the collapsed buildings, looking for clues. They interviewed hundreds of witnesses and persons involved in the design of the towers. Team members culled through thousands of pictures and hours of video. They examined how the building was constructed and how it was maintained. The information gathered will no doubt prove helpful for future design and construction of high-rise buildings.

Of course, the valuable information gathered from FEMA's research must be applied by state and local governments as well as private industry. This often happens when building codes are revised to reflect the findings of the research. In an effort to transfer this knowledge, FEMA distributes their findings via website, workshops, technical manuals, and training through the Emergency Management Institute (EMI) at the National Emergency Training Center.

Over the years, BPAT has experienced many success stories. One example of successful application of lessons learned is the Mobile, Alabama, Convention Center. Throughout history, Alabama has been susceptible to hurricanes and tropical storms. The city of Mobile desired to build a structure that would be hurricane resistant, yet aesthetically pleasing and cost effective. In addition to strong winds, hurricanes often bring floodwaters, so the convention center was built above the 100-year flood elevation.

In 1998, Hurricane Georges swept through the southern United States coastline, deluging Mobile with nearly 14 inches of water. Large portions of the downtown Mobile area were flooded. The Convention Center only received minor damage from the hurricane, and most of it was water related. The lower level of the building, which was primarily used for parking, was covered with approximately four feet of water. Because of this, some electrical repairs were necessary in addition to repair of the fire control panel and electronic parking equipment. Minor drywall repair and general removal of mud and water was also necessary. Total costs for repairs were approximately $350,000—a very small percentage of the cost of the $52 million structure.

Thanks to a design that addressed the possibility of hurricane induced floodwaters, the buildings HVAC equipment was elevated and remained unharmed during the hurricane. If this one simple design had not been featured, the repairs to the Convention Center could have cost millions of dollars.

Because the damage to the Convention Center was minor, the center was able to remain in operation and no events were canceled. In fact, the Convention Center was able to book additional events from other local facilities that were affected by the hurricane. If the Convention Center were not available shortly after the storm, a major convention of the Southeast United States/Japan Trade Commission Conference would have been canceled, costing more than $1 million in revenue. It is estimated that the design and elevation of the building represented a major economic success that exceed $5 million dollars in terms of both damages avoided and lack of business interruption costs. This is the type of success for which FEMA's Building Performance Assessment Team program strives.

Summary

Time and again history proves to be a great teacher. Much has been learned when it comes to understanding emergency events and how to limit their negative affects. Although we cannot eliminate many emergencies, we can change how we prepare for them and how we react to them when they do occur. We can also go back after the emergency has occurred and examine the affected buildings. Did the building survive? If it did, why did it survive? By analyzing the affect the emergency event had on the building, we can begin to develop better building codes. We can specify better building materials. We can take the lessons that history has taught us and design better buildings. Doing so will advance business continuity, reduce insurance costs, and—most importantly—save lives.

Appendix

Emergency Action Plan.. 213
    Buildings Covered.. 214
    Copies if this Manual Assigned to.. 214
    The Mission Statement.. 215
    The Emergency Management Team.. 216
    Chain of Command... 217
    Vendors and Contractors... 218
    Figure A-1. Contractor Availability Form.. 218
    Emergency Supplies.. 219
    Figure A-2. Storm Preparations List... 220
    Blueprints and As-Builts.. 222
    Insurance... 223
    Drills, Table Top Exercises, and Inspections... 224
    Actual Evacuation... 225
    Evacuation Script.. 225
    The Staging Area... 226
    Emergency First Aid Kit.. 227
    Generator Maintenance... 229
    Alarm Systems.. 230
    The Command Center.. 231
    Data Security... 232

Off-site Storage... 232

Natural Events... 233

Non-Natural Events... 241

Post-Event Restoration.. 248

Sample Press Releases... 250

OSHA Record Keeping... 251

Figure A-3. Pre-Storm Activities List.. 252

Figure A-4. Post-Storm Activities List... 253

Administrative Issues Annual Audit.. 254

Emergency Plan Issues Annual Audit... 256

General Building Issues Annual Audit.. 261

Health Related Issues Annual Audit... 270

Specific Threat Annual Audit... 271

Figure A-5. ATF Bomb Search & Sweeping Procedures..................................... 273

Figure A-6. Risk Assessment Matrix.. 276

Figure A-7. Correspondence Log.. 277

Figure A-8. OSHA 300 Recording Forms and Instructions................................. 278

Helpful Websites... 290

Where Else Can I Find Assistance?... 290

Here are just a few of the companies that manufacture the NOAA weather radio........ 298

Emergency Action Plan

for

[Insert Your Organization Name Here]

[Insert Your Address Here]

[Insert Your Company Logo Here]

Date of Current Revision

[Insert Date Here]

Revision

[Insert Revision Number Here]

Next Review Date

[Insert Date Here]

Buildings Covered

[Insert Building Names / Addresses Here]

Copies Of This Manual Assigned To:

[Insert Names / Titles Here]

The Mission Statement

[Add a copy of your emergency management team's mission statement here]

For Example:

Emergency Management Team – Mission Statement

Our mission is to protect the employees, property, business, and community in the event of an emergency.

We are committed to providing the company with effective and useful solutions that will reduce or eliminate emergencies and their negative impact.

We will accomplish this by conducting risk assessments, creating a written emergency action plan, and conducting employee training.

Additionally, we will create viable plans for restoration and recovery in the event of an emergency.

The Emergency Management Team

Title	Phone	Cell	Work Location
Emergency Director			
Business Continuity Manager			
Risk Management Manager			
Logistics Manager			
Procurement Manager			
Media Relations			
Human Resources			
Facilities Manager			
Legal Counsel			
Command Post Leader			
Group Leader			
Floor Captains			
Fire Brigade			
Evacuation Assistants			
Runners			
Other Important Contacts			
In-house medical contact (CPR, First Aid)			
Police Department			
Fire Department			
Health Department			
Poison Control Center			
Hospital			

Chain of Command

In the event of an emergency *[Insert your answer here]* will be in command. In the event this person is unavailable, *[Insert your answer here]* will take command.

Vendors and Contractors

During an actual emergency, we will likely need the resources of our vendors and contractors. In an effort to identify those vendors and contractors, we have created a database including the specialties of each vendor. That list can be found below.

Because this database can become outdated, we will review the list each year on *[insert your answer here]*.

Contractor Availability Form: First Alternate

Issue	Contractor	Contact	Office Phone	Cell Phone	AVAILABLE	NOT AVAILABLE	LEFT MESSAGE	Comments
Generator								
Landscaper								
Sign Maker								
Parking Lot Sweeper								
Barricade Supplier								
Tool Rental								
Security Guard Service								
Motorized Gate								
Fence Contractor								
Roofing Contractor								
Ice Machine								
Ice Supplier								
Dumpster Rental								
Refuse Company								
Trash Hauler								
Port-O-Pottie Supplier								
Computer Systems								
Computer Network Tech								
Phone Technician								
HVAC Contractor								
Electrical Contractor								
Engineering								
Architect								
Plumber								
Locksmith								
Glass Company								
Movers								
Carpet Cleaner								
Janitorial								
Laundry Services								
Elevator Service								
Fire Protection Vendor								
Security System								
Electric Utility								
Gas Utility								
Water Utility								
Phone Company								

Figure A-1. Contractor Availability Form

Emergency Supplies

The stocking of emergency supplies is an integral part of our emergency planning responsibilities. We also realize that the rotation of many of these supplies is critical. Therefore, we will review our emergency stock of supplies on an annual basis. This review will occur each year on *[insert your answer here]*.

To better facilitate the emergency stock of supplies, we have created a database which identifies materials to be stocked, the quantity needed, and the location where stored.

STORM PREPARATIONS
Inventory of Materials

	Location A	Location B	Location C	Location D
TOOLS / EQUIPMENT				
Chain saws (18') with 1 gallon gas can	1	1	2	1
1 quart bar & chain oil	2	2	4	2
2-cycle engine oil	2	2	4	2
Shovels	4	4	4	4
Wet vacs (25 gallon)	2	2	2	2
Fire estinguisher	4	4	6	4
Axe	1	1	2	1
Crowbar	1	1	1	1
Push Brooms	2	2	2	2
Chains and locks	2	2	2	3
Extension cords	3	3	4	3
Circular Saw	1	1	1	1
Screw guns / drills	2	2	2	2
Duct Tape	3	3	3	3
60 mil plastic liner (roll)	2	2	2	2
Orange spray paint	1	1	1	1
Ladder	1	1	1	1
Fans	2	2	4	2
Barricades	4	4	8	4
Emergency caution tape	2	2	2	2
FUEL SUPPLIES				
Gas can (5 gallon)				
Propane gas (Bottles for lanterns)				
Propane - 20lb Bottle for gas grill				
LIGHTING				
Propane lanterns				
6-volt lanterns				
Flashlights				
Portable quartz lighting				
BEDDING				
Folding cots				
Wool blankets				
CLOTHING				
Rubber boots, Size 10				
Rubber boots, Size 11				
Rubber boots, Size 12				
Work gloves				
Rain jackets, large				
Rain hoods				
Rain pants				
Hard hats				
Dust masks				
Respirators				
Safety glasses/goggles				
Reflective vests				
ROPE/WIRE				
Tie wire (50' Roll)				
Rope, 1/2" (300' Roll)				
Rope, 3/8" (200' Roll)				
PUMPS				
Sump pump w/hose and discharge hose				
MEDICAL				
First Aid kits				
Aspirin				
Sunblock				
Insect repellent				
WATER HOSES				
2 1/2" 6' Hose pkgs.				
SAND SUPPLIES				
Burlap bags				
Sand delivered (by load)				
GENERATORS				
Portable generator				
Fuel (unleadead gasoline)				
Oil (for Diesel Generator)				

Figure A-2. Storm Preparations List

STORM PREPARATIONS
Inventory of Materials

	Location A	Location B	Location C	Location D
RADIOS / BATTERIES				
Radios, NOAA				
Radios, 2-Way				
Clock (battery operated)				
Batteries - (size AA to D and 6V lantern)				
Calculator				
WATER / ICE / FOOD				
Ice chests (50 qt)				
Ice				
Meals Ready To Eat (MRE's)				
Matches				
Utinsils				
Garbage bags				
Water purifying tablets				
Bottled water - 5 gal. bottles				
Can opener (manual)				
BUILDING MATERIALS				
Plywood (4' x 8' x 5/8" sheets)				
2"x4"x 8' studs				
Screws				
Nails				
RENTAL EQUIPMENT				
Port-O-Potties				
Boat w/oars and accessories				
PHOTO EQUIPMENT				
Digital camera				
Disposable camera (Waterproof)				
Video				
CASH				
Emergency cash				
OFFICE SUPPLIES				
Steno pads				
Highlighters				
Pens				
Pencils				
Clipboards				
Paper clips				
Staples				

Figure A-2. Storm Preparations List, continued

Blueprints and As-Builts

Quick access to current copies of the floorplans of our buildings is important to our emergency management goals. To achieve this goal we have included updated floorplans as part of this emergency action plan. We have identified the following import building features on the floorplans.

- Windows
- Doors
- Emergency exits
- Alarm panels
- Fire alarm pull stations
- First aid kits and stations
- Automated External Defibrillators (AED)
- Generator location
- Generator-powered circuits
- Uninterruptible power system location
- Evacuation staging areas
- Gas mains and valves
- Utility shutoffs
- Generator and switchgear locations
- Water lines and valves
- Storm drains
- Roof drains
- Sewer lines
- Floor plans
- Alarm locations
- Water sprinkler testing valves
- Fire extinguishers
- Fire pull stations
- Smoke detectors
- Designated escape routes
- Exits doors
- Stairways
- Restricted areas
- Mechanical Rooms
- Telecommunications Rooms
- Haz-Mat storage locations

[Choose an answer]

We have included copies of our current floorplans in this emergency action plan.

Or

We keep copies of current floorplans at [Insert location here].

Insurance

We understand that proper insurance coverage is vital to our organization's survival in the event of an emergency. To ensure that we are properly insured, we conduct an annual review of our policies. This annual review was last conducted on *[Insert your answer here]*. Our next annual review will occur on *[Insert your answer here]*. The annual review will be conducted by *[Insert your answer here]*.

Drills, Table Top Exercises, and Inspections

Date of last evacuation drill [Insert your answer here]

    ? Partial   (Which areas) [Insert your answer here]

    ? Full

Scheduled frequency [Insert your answer here]

Next scheduled evacuation drill [Insert your answer here]

Date of last tabletop exercise [Insert your answer here]

Scheduled frequency [Insert your answer here]

Attended by [Insert your answer here]

Topics discussed [Insert your answer here]

Next scheduled table top exercise [Insert your answer here]

Date of last inspection by fire department [Insert your answer here]

Scheduled frequency [Insert your answer here]

Attended by [Insert your answer here]

Next scheduled fire department inspection [Insert your answer here]

Actual Evacuation

In the event of an emergency, it may be necessary to evacuate the building. If this happens, the evacuation command will be given by *[Insert your answer here]*. This evacuation command will be communicated via *[Insert your answer here]*.

The following employees are assigned as floor captains:

[Insert your answer here]

The following employees are assigned to assist those with disabilities:

[Insert your answer here]

The following employees are trained for fire extinguisher use:

[Insert your answer here]

The following employees are responsible for equipment shutdown:

[Insert your answer here]

The following employee is responsible for bringing the MSDS logbook and delivering it to fire rescue personnel:

[Insert your answer here]

Our procedures for accounting for employees after evacuation is [Insert your answer here]

In the event of the following emergency events, evacuation will not be initiated:
- Civil disorder
- Chemical spill outside of the facility
- Severe weather
- Earthquake
- Other *[Insert your answer here]*

Evacuation Script

"We have received an alarm in _____ building. As a precaution, we are requiring evacuation of all employees in _____. Please immediately take your personnel belongings and proceed in an orderly fashion to your assigned exterior staging areas."

The Staging Area

In the event of an evacuation, employees and visitors are instructed to gather in their pre-assigned staging areas. A site plan which identifies the various staging areas is included in this plan (see the following).

The following employee will be assigned as Command Post Leader:

[Insert your answer here]

In the event of the assigned Command Post Leader's unavailability, the following employee will be assigned as a backup Command Post Leader:

[Insert your answer here]

The following employees will be assigned as Group Leaders for the pre-assigned staging areas:

[Insert your answer here]

The following employees will be assigned as Runners between the pre-assigned staging areas and the Command Post leader:

[Insert your answer here]

[Insert copy of site plan with staging areas identified here]

Emergency First Aid Kit

We keep our first aid kits in the following location(s): *[Insert your answer here]*

The following employee is responsible for regular first aid kit evaluation and restocking: *[Insert your answer here]*

Review and restocking of first aid kits are performed *[how often – insert your answer here]*

The following employees have been trained in first aid administration:

[Insert your answer here]

The following employees have been trained in CPR:

[Insert your answer here]

The following employees have been trained in AED usage:

[Insert your answer here]

The first aid kit is an integral part of our emergency action plan. We have created a first aid inventory list.

[Add or delete contents per your requirements]
- Adhesive tape
- Adhesive strips (¾" x 3" and 1" x 3")
- Ammonia inhalants
- Antacid tablets
- Antiseptic ointment or wipes
- Aspirin tablets
- Blanket
- Biohazard bags
- Cold pack
- Cotton tip applicators
- Eyewash
- Fingertip bandages
- Gauze pads and roller gauze (assorted sizes)
- Hand cleaner
- Ibuprofen tablets
- Knuckle bandages
- Latex gloves
- Non-aspirin tablets
- Notepad and pen (actions and inventory)
- Plastic bags
- Scissors and tweezers
- Small flashlight and extra batteries
- Triangular bandage

- Activated Charcoal (use only if instructed by Poison Control Center)
- Assorted splints
- Blood pressure kit
- Burn relief spray
- CPR mask with oxygen port
- Defribulator
- Insect bite relief spray
- Insta-Glucose
- Nitrile gloves
- Stethoscope
- Syrup of Ipecac (use only if instructed by Poison Control Center)

Generator Maintenance

An emergency power generator is an important part of our emergency action plan. It must be properly maintained, tested, and fueled. To ensure this happens, the following procedures have been established:

[If your generator does not include an automatic transfer switch]

The following employee(s) will be responsible for switching generator power on:

[Insert your answer here]

The following employee(s) will confirm the generator power has transferred properly: *[Insert your answer here]*

In the event emergency generator power does not transfer properly or the generator does not start, *[Insert your answer here]*, would be notified.

[Insert your answer here] is the contractor assigned to provide regularly scheduled maintenance on the generator.

[Insert your answer here] is responsible for checking and ordering fuel for the generator.

[Insert your answer here] is the vendor that provides generator fuel delivery. Our generator(s) uses *[Insert your answer here – diesel, gasoline, etc.]* fuel.

Alarm Systems

Alarms are an important part of our emergency action plan. We have identified the following alarms in our buildings:

[Insert your answer here; location/type]

[Insert your answer here; location/type]

[Insert your answer here; location/type]

Regular inspections of these alarms are performed by *[Insert your answer here]*

The Command Center

In the event of a major emergency, we have established a central command center for our organization at *[Insert your answer here]*.

In the event of a small-scale emergency, we may elect to set up a command center onsite. The employee (and title) responsible for making this call is *[Insert your answer here]*. That employee can be reached at *[Insert your answer here]*.

Data Security

Data security and integrity is vital to the survivability of our organization.

Our organization's contact person for data security is *[Insert your answer here]*.

Procedures for data security include *[Insert your answer here]*.

We protect against virus infection by *[Insert your answer here]*.

We protect against hackers by *[Insert your answer here]*.

Off-site Storage

Due to the critical nature of our organization's electronic and hard copy data, we have elected to use off-site storage of data. The company we have contracted with for off-site storage is *[Insert your answer here]*. Their address and contact information is *[Insert your answer here]*.

Our organization's contact person for off-site storage is *[Insert your answer here]*.

Procedures for off-site storage of data includes *[Insert your answer here]*.

Natural Events

As part of our organization's emergency action plan, we have developed certain procedures for a number of different natural and weather-related events. These procedures are noted below.

Blizzards and Avalanches

We have determined that severe winter weather could affect our organization; therefore, we have elected to include the procedures below.

During inclement winter weather, we keep a rotating schedule for maintenance employees to be on call after hours in the event of an emergency. This schedule and the direction of work is the responsibility of *[Insert your answer here]*.

Included in this schedule is:
- Snow removal and salting of walkways.
- Check roof drains, gutters, catch basins, and other areas that can freeze or get clogged and impede the flow of melted snow.
- *[Other - Insert your answer here]*.

Winterizing of HVAC, fire sprinkler, and irrigation systems will occur during the fall. The person responsible for scheduling and directing this task is *[Insert your answer here]*.

In the event of inclement weather that forces shut down of the building, the following people will remain on site:

[Insert your answer here].

[Insert your answer here].

The decision to have certain employees remain on-site is the responsibility of *[Insert your answer here]*.

Drought

We have identified a list of specific locations that are drought prone. These areas include properties that have areas of vegetation exist, border on wooded areas, are non-irrigated or high visibility areas. These locations include

[Insert properties and addresses here]

[Insert properties and addresses here]

[Insert properties and addresses here]

We have established internal terms that identify the severity of the drought. These terms include Advisory, Alert, Emergency, and Ration.

The announcement of each of these conditions will result in the following actions:

<u>Advisory</u> – When an advisory notice is given we will make an announcement to employees. The condition may escalate within the next five days, so we will review our procedures for each condition. Other steps include *[Insert your answer here]*.

Alert – When an alert order is given we will announce a voluntary reduction of water usage to employees and perform visual inspections of the premises to identify fire hazards. Other steps include *[Insert your answer here]*.

Emergency – When an emergency order is given, we will cut back vegetation and release a memo to employees and contractors concerning potential fire hazards. Specific care will be taken when welding or roofing work is being performed. Other steps include *[Insert your answer here]*.

Ration – When a ration order is given, water usage will be decreased at the direction of the local authorities. All irrigation on automatic timers will be adjusted accordingly. Other steps include *[Insert your answer here]*.

Earthquakes

We have determined that earthquakes are a possibility in our area; therefore, we have developed a number of earthquake specific safety procedures:

- In the event of an earthquake we have identified the following safe locations in the buildings:

  [Insert your answer here].

  [Insert your answer here].

  [Insert your answer here – you may elect to insert of copy of your floorplan here].

- Conversely, we have identified a number of dangerous areas in our buildings during an earthquake:

  [Insert your answer here].

  [Insert your answer here].

  [Insert your answer here – you may elect to insert of copy of your floorplan here].

- In the event of an earthquake, we understand that it may be advisable to turn off certain utilities. The employee responsible for assessing the utilities and making a decision to shut down is *[Insert your answer here]*.

We have trained employees to take refuge beneath a desk or table. If that is not possible, employees should move toward an inside wall away from items that could fall and injure them such as bookcases, vending machines, lighting, and other fixtures.

If employees are outdoors, they should move away from buildings, radio towers, light poles and overhead utility lines. In a high-rise building, do not use elevators.

In the event of an earthquake that causes structural damages, partial structural collapse, or total collapse, care must be taken to ensure that employees, contractors, and other visitors to the site are safe.

If an earthquake occurs that causes structural damage or collapse to a portion of the building, the facility manager must perform a building assessment. If the facility manager does not have the expertise to perform this assessment or if the scope of the assessment exceeds the expertise of the facility manager, we will use an architectural (or engineering) firm to complete this assessment. Our approved contractor for this assessment is *[Insert your answer here]*.

If the earthquake causes damage to the building, care should be taken to ensure the safe operation of utilities. If utility lines have been compromised, they must be turned off. The employee(s) responsible for utility shut off include *[Insert your answer here]*.

When entering a building that has experienced any level of collapse or structural damage, personal protective equipment (PPE) must be issued to employees and worn by employees and contractors at all times. A PPE station shall be erected on site and a list of required PPE for entry shall be posted. The employee responsible for this station and PPE distribution is *[Insert your answer here]*.

Fire

Fire could occur in any building or location; therefore, fire protection is an important part of our emergency action plan. We have developed a list of all the major workplace fire hazards and their proper handling and storage requirements. These hazards include:

[insert your answer here - i.e. welding, smoking]

[Insert your answer here].

Their safe handling and storage requirements include:

[Insert your answer here].

[Insert your answer here].

Housekeeping procedures are important to ensure that work areas are kept free from accumulations of flammable and combustible materials. Housekeeping procedures include:

[Insert your answer here].

[Insert your answer here].

We have identified the following types of fire protection available to control each hazard:

[Insert your answer here].

[Insert your answer here].

The names of personnel responsible for maintenance of equipment installed to prevent or control fires include:

[Insert your answer here].

[Insert your answer here].

We understand that maintenance of fire equipment or systems (i.e., inspections, certifications) is an integral part of our emergency action plan. Our maintenance is conducted by *[Insert your answer here]*. The frequency of such maintenance is [Insert your answer here].

A water sprinkler system is the most effective method of controlling fires automatically. We use the following types of fire sprinkler systems: *[insert your answer here – i.e., Wet Pipe System, Dry Pipe System, or Deluge System]*

In the event of a fire that causes structural damages, partial structural collapse, or total collapse, care must be taken to ensure that employees, contractors, and other visitors to the site are safe.

If a fire occurs that causes structural damage or collapse to a portion of the building, the facility manager must perform a building assessment. If the facility manager does not have the expertise to perform this assessment or if the scope of the assessment exceeds the expertise of the facility manager, we will use an architectural (or engineering) firm to complete this assessment. Our approved contractor for this assessment is [Insert your answer here].

If the fire causes damage to the building, care should be taken to ensure the safe operation of utilities. If utility lines have been compromised, they must be turned off. The employee(s) responsible for utility shut off include [Insert your answer here].

When entering a building that has experienced any level of collapse or structural damage, personal protective equipment (PPE) must be issued to employees and worn by employees and contractors at all times. A PPE station shall be erected on site and a list of required PPE for entry shall be posted. The employee responsible for this station and PPE distribution is [Insert your answer here].

Floods

We have determined that floods can occur in our area and could negatively affect our organization. We cannot always control floodwaters, but we can keep adequate flood insurance coverage. The employee responsible for insurance coverage is [Insert your answer here]. This employee will review our insurance coverage [Frequency - Insert your answer here].

We also believe that a proactive approach to floods is important; therefore, we have developed the following procedures:

[Insert your answer here – i.e., disconnect appliances and equipment as water rises].

[Insert your answer here – i.e., keep a supply of sand, sandbags, lumber etc].

[Insert your answer here – i.e., move valuables to higher area].

[Insert your answer here – i.e., use water pumps to displace water].

[Insert your answer here – i.e., dehumidify the building after water has receded].

[If you are located near a river and within fifty miles of a dam, explain any special considerations or communication with the agency responsible for the dam operations.]

In the event of a flood that causes structural damage, care must be taken to ensure that employees, contractors, and other visitors to the site are safe.

If a flood occurs that causes structural or cosmetic damage to a portion of the building, the facility manager must perform a building assessment. If the facility manager does not have the expertise to perform this assessment or if the scope of the assessment exceeds the expertise of the facility manager, we will use an architectural (or engineering) firm to complete this assessment. Our approved contractor for this assessment is [Insert your answer here].

If the flood causes damage to the building, care should be taken to ensure the safe operation of utilities. If utility lines have been compromised, they must be turned off. The employee(s) responsible for utility shut off include [Insert your answer here].

Hurricanes

We have established specific procedures for arriving storms. These procedures address actions taken 72 hours prior to the storm's arrival, 36 hours prior to the storm's arrival, and 24 hours prior to the storm's arrival. These activities are identified on the Pre-storm List found below.

[Attach Pre-storm List here]

Additionally, specific procedures to be taken after the storm has passed are found on the attached Post-storm List.

[Attach Post-storm List here]

Tropical Storm Watch

Within the next 36 hours, tropical storm conditions (winds from 36 to 73 mph) are possible in the storm watch area.

Tropical Storm Warning

Within the next 24 hours, tropical storm conditions (winds from 36 to 73 mph) are expected in the storm warning area.

Hurricane Watch

Within the next 36 hours, hurricane conditions (sustained winds greater than 73 mph) are possible in the hurricane watch area.

Hurricane Warning

Within the next 24 hours, hurricane conditions (sustained winds greater than 73 mph) are expected in the hurricane warning area.

Additionally, the following related warning might be given:

Flood Watch

Floodwaters are possible in the given area of the watch and during the time identified in the flood watch. Flood watches are usually issued for flooding that is expected to occur at least 6 hours after heavy rains have ended.

Flood Warning

Flooding is actually occurring or is imminent in the warning area.

Flash Flood Watch

Flash Flooding is possible in the warning area. Flash flood watches are usually issued for flooding that is expected to occur within six hours after heavy rains have ended.

Flash Flood Warning

Flash flooding is actually occurring or is imminent in the warning area.

Saffir-Simpson Rating and Storm Categories:

1. **Minimal:** Winds of 74 to 95 mph. Damage is minimal, usually limited to trees and power lines.

2. **Moderate:** Winds of 96 to 110 mph. Some roof damage may occur. Trees can be uprooted and power poles can be downed. Windows and storefronts can be damaged.

3. **Extensive:** Winds of 111 to 130 mph. Roofs are badly damaged or lost, structural damage is common.

4. **Extreme:** Winds of 131 to 155 mph. Damage to most structures is severe. Extreme flooding can occur along and near the coast. Loss of life is common.

5. **Catastrophic:** Winds in excess of 155 mph. Damage is total in many places, and many structures are destroyed. Severe flooding is common several miles inland.

In the event of a hurricane that causes structural damage, partial structural collapse, or total collapse, care must be taken to ensure that employees, contractors, and other visitors to the site are safe.

If a hurricane occurs that causes structural damage or collapse to a portion of the building, the facility manager must perform a building assessment. If the facility manager does not have the expertise to perform this assessment or if the scope of the assessment exceeds the expertise of the facility manager, we will use an architectural (or engineering) firm to complete this assessment. Our approved contractor for this assessment is *[Insert your answer here]*.

If the hurricane causes damage to the building, care should be taken to ensure the safe operation of utilities. If utility lines have been compromised, they must be turned off. The employee(s) responsible for utility shut off include *[Insert your answer here]*.

When entering a building that has experienced any level of collapse or structural damage, personal protective equipment (PPE) must be issued to employees and worn by employees and contractors at all times. A PPE station shall be erected on site and a list of required PPE for entry shall be posted. The employee responsible for this station and PPE distribution is *[Insert your answer here]*.

Tornadoes

We have determined that tornados could occur in locations that we have properties; therefore, we have established a number of tornado-specific emergency procedures.

If a tornado is spotted, we have trained our employees to immediately contact 911 to report it. We have also trained employees to contact *[Insert your answer here – internal company contact]*.

In the event of a tornado strike, employees have been instructed to stay away from windows and doors. *[Does your building have a basement? – If yes: Move to the basement. If a basement is not available, employees have been trained and will be instructed to go to a safe area of the building]*.

We have identified the following interior locations as safe areas:

[Insert your answer here]

Additionally, in the event of a tornado, all employees have been instructed to stay on the lowest level of the building.

We have a NOAA weather radio available and will stay tuned for further instructions.

Tornado Watch

Tornados are possible in the given area of the watch.

Tornado Warning

A tornado has actually been sighted by a spotter or by radar and is occurring or is imminent in the warning area.

In the event of a tornado that causes structural damage, partial structural collapse, or total collapse, care must be taken to ensure that employees, contractors, and other visitors to the site are safe.

If a tornado occurs that causes structural damage or collapse to a portion of the building, the facility manager must perform a building assessment. If the facility manager does not have the expertise to perform this assessment or if the scope of the assessment exceeds the expertise of the facility manager, we will use an architectural (or engineering) firm to complete this assessment. Our approved contractor for this assessment is *[Insert your answer here]*.

If the tornado causes damage to the building, care should be taken to ensure the safe operation of utilities. If utility lines have been compromised, they must be turned off. The employee(s) responsible for utility shut off include *[Insert your answer here]*.

When entering a building that has experienced any level of collapse or structural damage, personal protective equipment (PPE) must be issued to employees and worn by employees and contractors at all times. A PPE station shall be erected on site and a list of required PPE for entry shall be posted. The employee responsible for this station and PPE distribution is *[Insert your answer here]*.

Tsunamis

We have identified the threat of a tsunami as being a potential in our area. For this reason we have developed a number of procedures.

We have identified an elevated evacuation route. Our elevated area of refuge is at least 50 feet above sea level and one mile in from the coast. This area of refuge is *[Insert your answer here – location and directions]*

Because evacuation routes may be blocked or unavailable, we have identified a second route. This second area of refuge is *[Insert your answer here – location and directions]*

In the event of a tsunami that causes structural damage, partial structural collapse, or total collapse, care must be taken to ensure that employees, contractors, and other visitors to the site are safe.

If a tsunami occurs that causes structural damage or collapse to a portion of the building, the facility manager must perform a building assessment. If the facility manager does not have the expertise to perform this assessment or if the scope of the assessment exceeds the expertise of the facility manager, we will use an architectural (or engineering) firm to complete this assessment. Our approved contractor for this assessment is *[Insert your answer here]*.

If the tsunami causes damage to the building, care should be taken to ensure the safe operation of utilities. If utility lines have been compromised, they must be turned off. The employee(s) responsible for utility shut off include *[Insert your answer here]*.

When entering a building that has experienced any level of collapse or structural damage, personal protective equipment (PPE) must be issued to employees and worn by employees and contractors at all times. A PPE station shall be erected on site and a list of required PPE for entry shall be posted. The employee responsible for this station and PPE distribution is *[Insert your answer here]*.

Volcanoes

We have determined our organization has property in the vicinity of volcanoes. Those properties include:

[Insert your answer here]

In the event of a volcanic eruption we will immediately evacuate the area.

We have identified an elevated evacuation route. This area of refuge is *[Insert your answer here – location and directions]*

Because evacuation routes may be blocked or unavailable we have identified a second route. This second area of refuge is *[Insert your answer here – location and directions]*

In the event of a volcanic disturbance that causes structural damage, partial structural collapse, or total collapse, care must be taken to ensure that employees, contractors, and other visitors to the site are safe.

If a volcanic disturbance occurs that causes structural damage or collapse to a portion of the building, the facility manager must perform a building assessment. If the facility manager does not have the expertise to perform this assessment or if the scope of the assessment exceeds the expertise of the facility manager, we will use an architectural (or engineering) firm to complete this assessment. Our approved contractor for this assessment is *[Insert your answer here]*.

If the volcanic disturbance causes damage to the building, care should be taken to ensure the safe operation of utilities. If utility lines have been compromised, they must be turned off. The employee(s) responsible for utility shut off include *[Insert your answer here]*.

During initial return to the building, employees and contractors must be aware of the danger of soot, dust, and volcanic ash. When entering a building that has experienced any level of collapse or structural damage, personal protective equipment (PPE) must be issued to employees and worn by employees and contractors at all times. A PPE station shall be erected on site and a list of required PPE for entry shall be posted. The employee responsible for this station and PPE distribution is *[Insert your answer here]*.

Non-Natural Events

We understand that weather-related and other natural events are not the only events to be concerned with. A number of non-natural events could also occur, so we have elected to address them here.

We have identified the following potential hazards in the workplace:

[Insert your answers here – i.e., chemical spill]

[Insert your answers here – i.e., fire from welding]

[Insert your answers here – i.e., injury from machinery]

Bomb Threat

If faced with a bomb threat we have three basic options: Ignore the threat, evacuate immediately, or search for the bomb and evacuate if necessary. Our decision as to which option we select will depend on a number of factors including:

Ignore the threat – The threat lacks credibility. *[Insert your answer here; i.e., the threat lacks credibility]*

Evacuate immediately – The threat is very credible. A bomb or other device has been located. *[Insert your answer here; i.e., the threat is very credible]*

Search for the bomb and evacuate if necessary – The threat seems credible. *[Insert your answer here; i.e., the threat seems credible]*

[Insert your answer here] is the employee responsible for calling an evacuation if the local authorities elect not to get involved.

We have contacted local police and fire departments to determine if a bomb squad is available to assist us with search and removal of a bomb. Their process for assistance is *[Insert your answer here]*. The names and phone numbers of the local authorities are *[Insert your answer here]*.

We have designated search team members and they include:

[Insert your answers here].

Because of their familiarity with the area, a volunteer from each department or area being searched is also required. A set of floor plans will be available for the search team and authorities. This floorplan is kept at *[Insert your answer here]*

In the event of a bomb threat and evacuation we will establish a Command Center outside of the building. That exact location is *[Insert your answer here]*.

In the event a bomb or other suspicious device is found, *[Insert your answer here]* will be responsible for communicating our findings to the local authorities and asking for assistance.

Loss of Utility

An extended loss of utility could adversely affect our organization. Loss of utility might include electricity, water, or sewer. We have determined that loss of electricity could have the greatest impact on our business. This loss is especially critical following another emergency event because it would slow our ability to respond and recover.

We have identified the following critical areas of our business that would be most affected by loss of utility:

[Insert your answer here – i.e., production area]

[Insert your answer here – i.e., computer centers]

[Insert your answer here – i.e., call center]

To ensure business continuity we have identified the following critical components and their availability.

- Generator: [Insert your answer here - i.e., on-site generator located at...]
- [Insert your answer here – i.e.. flatbed generator to be shipped to...]
- Port-O-Potties: [Insert your answer here – i.e., available from...]
- Bottled Water: [Insert your answer here – i.e., available from...]
- Other: [Insert your answer here]
- Other: [Insert your answer here]

Medical Emergencies

We have determined that a medical emergency could affect our organization and our employees in a negative way; therefore, we have elected to develop certain medical related procedures.

We have identified the following areas of concern that could cause a medical emergency:

[Insert your answer here – i.e., machinery]

[Insert your answer here – i.e., confined spaces]

[Insert your answer here – i.e., chemical usage area]

We have identified employees who have emergency medical response training:

[Insert your answer here – include name, work location, office phone, and cell]

[Insert your answer here – include name, work location, office phone, and cell]

[Insert your answer here – include name, work location, office phone, and cell]

Because of the potential for disease transmission while administering first aid, our emergency response employees have been properly trained and outfitted with devices that limit the potential for disease transmission avoidance.

We have identified the closest hospital facility as [Insert your answer here] and emergency response time is estimated to be [Insert your answer here - minutes].

In the event of a medical emergency we will call [Insert your answer here]. If an outside line is required we will [Insert your answer here – i.e., if you dial 911, must you first dial "9" to get an outside line?]

We have located a First Aid kit and AED's on our floorplan included in this emergency action plan.

Within eight hours after a medical emergency or illness, our organization will perform all required OSHA injury and illness recordkeeping. The person or department responsible for this is *[Insert your answer here]*.

Chemical and Biological Concerns

Although not likely, we feel that the threat of chemical or biological concerns must be addressed in our organization. This is especially true in the event we were to receive mail that was tainted with biological agents.

In the event of receipt of a suspicious package, the employee handling the package would immediately seal or cover the package.

Next, all employees would leave the room and if possible lock it. If locking the room is not possible, the employees will post a sign warning not to enter. The package and the area it is stored must be quarantined.

All employees that came in contact with the package or were in the general vicinity of the package must immediately wash their hands with soap and water.

A list of people who were in the area and possibly exposed will be developed and given to the authorities.

Next, the employees in the affected area would notify the building manager, local police department, and the CDC Emergency Response Hotline.

CDC website: http://www.bt.cdc.gov/EmContact/index.asp

Additionally, the CDC Emergency Response Hotline can be called 24 hours a day at (770) 488-7100

Other procedures would include: *[Insert your answer here]*

Chemical Spills and Contamination

There is a risk that a chemical spill could affect our organization. We have identified the following potential spill or contamination sources:

[Insert your answer here – external sources, i.e., highway, navigable waters, nearby plants]

[Insert your answer here – internal sources, i.e., chemical storage, manufacturing]

In the event of chemical spill, employees must take certain precautions such as building evacuation or instructions to shelter in place.

If a decision is made instructing employees to shelter in-place, employees must not attempt to leave the building. It has been determined that the chemical has been released and exposure is likely to occur if you leave the protective confines of the building. Additionally, the following steps should be taken:

- Close and lock all doors and windows
- Seal all openings around doors with wet towels.
- Close all fireplace chimney dampers.
- Turn off all HVAC equipment and close or seal all fresh air intakes.

- Seal off, using plastic visqueen and duct tape, all other openings such as vents and exhaust fans.
- Other steps: *[Insert your answer here]*

To reduce the possibility of an internal chemical spill or contamination, the following steps must be observed:

Ensure that all containers storing chemicals are properly labeled.

Verify that all chemicals are stored in their proper containers.

Ensure that chemicals with the potential of a negative reaction to another chemical are stored in separate areas.

Verify that all chemicals have a Material Safety Data Sheet (MSDS) available.

Keep the MSDS logbook in a central location available to all employees.

Train employees to properly handle and store chemicals. Furthermore, train employees to recognize and how to properly handle chemical spills. These training records are kept *[Insert your answer here]*.

In the event of a chemical spill, our procedures to communicate the spill to employees, management, and, if necessary, the proper local agencies are *[Insert your answer here]*.

Nuclear Threat and Exposure

The opportunity exists for nuclear exposure that will affect our organization. We have identified the following nuclear exposure sources:

[Insert your answer here]

In the event of nuclear exposure, a decision may be made to shelter in place. If a decision is made instructing employees to shelter in-place, employees must not attempt to leave the building. It has been determined that nuclear energy has been released and exposure is likely to occur if you leave the protective confines of the building. Additionally, the following steps should be taken:
- Close and lock all doors and windows
- Seal all openings around doors with wet towels.
- Close all fireplace chimney dampers
- Turn off all HVAC equipment and close or seal all fresh air intakes.
- Seal off, using plastic visqueen and duct tape, all other openings such as vents and exhaust fans.
- Other steps: *[Insert your answer here]*

If instructions to leave the building and evacuate to a safe area have been given, follow the instructions given by the authorities.

[The following text applies to those near a nuclear plant]

The Nuclear Regulatory Commission requires that nuclear power plants have a system for notifying the public in the event of an emergency. There exist four standard classifications of emergencies, including:

- Notification of Unusual Event—This is a non-specific warning and is the least serious of the four categories. The event poses no danger to employees or the public and no action (i.e., evacuation) is required of the public.
- Alert—This classification is declared when an event has occurred that could jeopardize the plants safety, but backup systems are in place. Emergency agencies are notified, but no action is required from the public.
- Site Area Emergency—This classification, is declared when an event has caused a major problem with the plants safety system and has progressed to a point where a release of radiation into the air or water is possible but would not exceed regulations instituted by the Environmental Protection Agency. No action is required by the public.
- General Emergency—This classification is the most serious and comes when the plants safety systems have been lost. Radiation could be released beyond the boundaries of the plant. Emergency sirens will sound and some people will be evacuated or instructed to remain sheltered in place. Listen for more specific instructions.

If we receive an alert, remember the following:
- A siren or tone alert does not necessarily mean we should evacuate. Listen to the television or radio for further instructions.
- Do not call 911. If this is a true alert, a special rumor control phone number will be provided.
- If we are instructed to shelter in place, be sure to close all doors, windows, chimney dampers, and turn off all HVAC equipment.

If we receive a warning and you are instructed to go inside, it is advisable to shower and change clothes. After removing your clothes and shoes, place them in a plastic bag and seal the bag.

Civil Disturbance and Demonstrations

We have determined that, although unlikely, it would be prudent to include the possibility of civil disturbance in our emergency action plan. Events that may trigger civil disturbance and, therefore, should be carefully watched when occurring in the general proximity of our properties might include
- Labor disputes
- Layoffs and downsizing
- Environmentally sensitive meetings or conferences
- Sporting events
- Political rallies
- Economic conferences
- Judicial decisions
- Music concerts
- Religious gatherings
- Biased racial or cultural events

In the event of civil disturbance, we may elect to take the following actions:
- Hiring temporary guard service

- Installing storm shutters
- Locking gates in the parking lot
- Lowering of security grating
- Removal of vehicles from the premises
- Removal of trash containers or other items that could thrown or set afire
- [Other – Insert your answer here].
- [Other – Insert your answer here].

The decision to take any of these actions is the responsibility of [Insert your answer here].

Structural Collapse

Although a structural collapse of one of our buildings is not likely, we feet it is prudent to include this potential event in our emergency action plan.

In the event of an event that causes structural damage, partial structural collapse, or total collapse, care must be taken to ensure that employees, contractors, and other visitors to the site are safe.

If an event occurs that causes structural damage or collapse to a portion of the building, the facility manager must perform a building assessment. If the facility manager does not have the expertise to perform this assessment or if the scope of the assessment exceeds the expertise of the facility manager, we will use an architectural (or engineering) firm to complete this assessment. Our approved contractor for this assessment is [Insert your answer here].

If the emergency event causes damage to the building, care should be taken to ensure the safe operation of utilities. If utility lines have been compromised, they must be turned off. The employee(s) responsible for utility shut off include [Insert your answer here].

When entering a building that has experienced any level of collapse, personal protective equipment (PPE) must be issued to employees and worn by employees and contractors at all times. A PPE station shall be erected on-site and a list of required PPE for entry shall be posted. The employee responsible for this station and PPE distribution is [Insert your answer here].

In the event of a partial or total collapse, we understand that asbestos fibers could be released; therefore, proper care must be taken. We have identified the following buildings that feature asbestos-containing materials:

[Insert your answer here]

[Insert your answer here]

[Insert your answer here]

Contractors and employees performing work at locations that could produce an asbestos exposure must be informed of the potential asbestos exposure and must take proper precautions when working in the area. These precautions include [Insert your answer here]. Our preferred vendor for asbestos testing and abatement is [Insert your answer here].

In the event of a total collapse that has trapped people, we will immediately contact the Federal Emergency Management Agency (FEMA) at [Insert your answer here].

Workplace Violence

Workplace violence is an unfortunate reality, and we believe our organization has a responsibility to protect our employees. To accomplish this, management is committed to providing a safe environment for employees to work.

As an organization we have identified certain prohibited behavior including:

[Insert your answer here – i.e., threatening language].

[Insert your answer here – i.e., damage to company property].

[Insert your answer here – i.e., physically assaulting another person].

[Insert your answer here – i.e., sexually harassing another person].

[Insert your answer here – i.e., bringing weapons on company property].

Because of the serious nature of workplace violence and the incidents that could lead to violence, we reserve the right to discipline employees. Such disciplinary actions might include verbal warning, written warning, suspension and termination of employment. The severity of the penalty will be based on the severity of the infraction. The procedures for investigating reported incidents might include *[Insert your answer here]*.

Post-Event Restoration

As soon as it is safe to enter the property affected by an emergency event, the facility manager or those assigned by the facility manager will assess the damage to the property. Based on this assessment, the company will begin restoration procedures.

Insurance and salvage decisions will need to be made. Employees responsible for making these decisions include *[Insert your answer here]*.

Those employees expected to report to the affected location must be self-sufficient for the first 48 hours *[or insert your answer here]*. These first responder employees are responsible for providing the following items:

Clothing
- Three changes of clothes
- Rain jacket
- Cold weather jacket
- Safety goggles
- Hard hat
- Sun glasses
- Two pairs of work shoes (steel toe preferred)
- Spare pair of glasses
- *[Other - Insert your answer here]*.

Health and Toiletries
- Soap
- Shampoo
- Wash cloth and towel
- Razor and shaving cream
- Toothpaste and toothbrush
- Male or female personal hygiene products.
- Deodorant
- Toilet paper
- Sun block
- Insect repellant
- Contact lens solution and spare lenses
- *[Other - Insert your answer here]*.

Miscellaneous
- NOAA weather radio
- Alarm clock
- Blankets
- Sleeping bag or cot
- Batteries
- Flashlight
- Pen or pencils

- Notepad and audit forms
- Camera (digital or film)
- Video camera
- Cell phone with automobile battery charger
- Drivers license
- Credit card
- Cash
- Airtight plastic container to keep items dry
- *[Other - Insert your answer here].*

Medicines
- Prescription medicines
- Antacids
- Athletes' foot spray
- Pain relief cream
- Aspirin and ibuprofen
- First aid kit
- *[Other - Insert your answer here].*

Food and Water
- Meals Ready to Eat (MRE)
- Snack bars
- Fruit
- Trail mix or nuts
- Bottled water (drinking and cleaning)
- Instant coffee
- Cooler with ice and ice pack
- Propane stove (optional)
- Propane fuel tank (extra)
- *[Other - Insert your answer here].*

Sample Press Releases

[This is just an example. Insert samples of your press releases here]

Dear Fellow Employee,

Our business was built on hard work and dedication from our employees. That need for hard work and dedication has never been as evident as it is today. Your efforts are greatly appreciated and I have no doubt we will persevere. We are faced with an excellent opportunity to show our resolve and character, and I invite you to show your best.

When faced with unusual events such as what we have just experienced, a few reminders are in order. We will keep our employees updated via e-mail and fax service. For those unable to access these mediums, we have established an emergency hotline (800-123-1234) with recorded updates and frequently asked questions. We realize that rumors will surface. We ask that you do not perpetuate these comments. All official communication will be provided through the methods listed above.

For many of our employees, the media will be observing your work. Again, you have an opportunity to present the company in a positive way. Unfortunately, some media representatives may attempt to sensationalize disasters and how they are dealt with. If asked about your restoration efforts, you may briefly explain what you are doing, stressing the fact that your efforts are part of a larger restoration effort by all of the employees of the company.

We ask that you refer questions involving detailed aspects of your work to our corporate communication department at (800-555-1212). Finally, if you don't know an answer to a reporter's question, don't speculate! It's ok to say, "I don't know." If you find yourself in an uncomfortable situation, thank the reporter for their time, and kindly remind them that you need to get back to work.

I am very proud of the efforts of our employees and I know we can all count on each other to make a complete recovery a reality. Remember, we have an opportunity to make a difference and your efforts are greatly appreciated.

President – Acme Industries

OSHA Record Keeping

Our organization is committed to employee safety. We are also obligated to OSHA compliance. With this in mind, especially after an emergency event, we have developed the following OSHA recordkeeping procedures:

[Insert your answer here – i.e. who is responsible for OSHA recordkeeping?].

[Insert your answer here – i.e. where are the OSHA recordkeeping forms kept?].

[Insert your answer here – i.e. who is responsible for annual posting of OSHA Form 300A?].

Within eight (8) hours after the death of any employee from a work-related incident or the in-patient hospitalization of three or more employees as a result of a work-related incident, you must orally report the fatality/multiple hospitalization by telephone or in person to the Area Office of the Occupational Safety and Health Administration (OSHA), U.S. Department of Labor, that is nearest to the site of the incident. You may also use the OSHA toll-free central telephone number, 1-800-321-OSHA (1-800-321-6742).

Storm Activities	(72 Hours) Before	(36 Hours) Before	(24 Hours) Before	Responsible Employee	Contractor/Company	Office Phone	Cell Phone
Grounds							
Shrubs and hedges	Survey/Trim as needed		Bring all planters into the building				
Trees	Trim all broken/dead limbs						
Light and flag poles			Remove all flags				
Newspaper stands		Bring into building or call newspaper vendor					
Drainage		Clear all drains of debris					
Parking		Make sure manual overide on gates are funtional	Remove vechicles from areas near trees				
Access							
Vehicle Fuel Tanks			Top off fuel tanks				
Building Exterior							
Shutters	Inventory all detachable shutters to verify completeness.		Close and secure all shutters				
Window Leaks		If appropriate seal leak. Indicate location on building floor plan. Move any materials/equipment away from leak area.	Same as 36Hrs before				
Rain Gutters		Remove debris from gutter	Remove debris from gutter				
Sand Bags	Notify sand supplier for delivery	Fill and place sand bags in low lying areas (doorways)					
Water Pumps	Motify rental company of impending need	Request pump(s) delivered. (250' - 500' hose)					
Building Interior							
Water System	Notify bottled water contractor of impending need.	Place order for 25 5-gallon water bottles					
Sewer System	Notify Port-o-Pottie of impending need	Request Port-o-Potties. (Store or have delivered post storm)					
A/C System	Inventory portable fans on hand	Rent any additional portable fans necessary					
Generator System	Notify generator maintenance contractor and top off fuel	Contractor check all operations. (Test WITH Building Load)	Contractor MUST have completed all checks and have generator operational				
UPS System	(Do preventative maintenance check prior to storm season)						
Telephone System	Notify all people on the storm call list. Verify all phone numbers on list	Make corrections to phone numbers when they are reported	Make corrections to phone numbers when they are reported				
Radio	Set up a radio distribution list. Test all radios and batteries. Repair/replace as needed.	Distribute all radios per distribution list	All radio operators check in				
Garbage Pickup/Removal	Call for a pickup for all Dumpsters						
Food Operations	Notify food service management. Inventory food stocks (for 2-day supply, 4 meals per day). Place order for shortages.	Notify food service staff	Food service manager contact departments to verify meal schedule				
Refrigerators			Set to a colder setting and limit access				
TV & Cable Systems	Check all monitors for proper channel and operations		If antennas used, take down				
Fire Alarm System	Update callout list with monitoring contractor		Remove locks on shut-off valves				
Security System	Check all electro/mechanical locks for proper function. Repair any problems. Notify security contractor to determine personnel stat.s	Meet with departments to determine the number of people that will be on site DURING storm	Security guards are briefed on their role and given specific instructions. Security contractor places all ards (incl. din. any additional				
Storm Supply Storage	Identify (on map) potential staging areas for supplies to be stored during the storm		Clear out staging area of any non-storm material				
Miscellaneous							
Critical documents	Identify critical documents		Pack documents and move 50 to 75 miles inland away from storm				
Insurance	Review policy. Take video of interior and exterior						
Inventory List	Complete and replinish where necessary						
Contractor Avaliability	Call contractors to ensure availability		Call key contractors again				

Figure A-3. Pre-Storm Activities List

Storm Activities	After	Responsible Employee	Contractor Company	Office Phone	Cell Phone
Grounds					
Shrubs and hedges	Survey site. Indicate all damage on site map. Report to landscape contractor				
Trees	Survey site. Indicate all damage on site map. Report to landscape contractor				
Light Poles & Flag Poles	Survey site. Indicate all damage on site map. Report to electrical contractor				
Newspaper stands	Return to original location				
Drainage	Unclog all drains				
Parking	Remove all debris. Indicate helicopter landing areas				
Access	Check all gates for proper operation				
Vehicle Fuel Tanks	Order fuel, if needed				
Building Exterior					
Shutters	Open/Remove all shutters. Store removable shutters in designated area. Open windows to air				
Window Leaks	Report all damage to Window contractor				
Rain Gutters	Remove all debris from gutter				
Sand Bags	Remove sand bags after water has subsided				
Water Pumps	Set-up and run until water subsides				
Building Interior					
Water System	Repair all damages to water system. Disperse water bottles. Replenish water bottles until water system				
Sewer System	Schedule pickups of Port-o-Potties until sewer system is operational				
A/C System	Repair all damage. (Check roof top units visually)				
Generator System	If activated, monitor operations. Check usage and refuel if necessary				
UPS System	Report all irregularities to UPS maintenance contractor				
Telephone System	Inform telecommunications contractor of any irregularities.				
Radio	Radio operators check in per schedule or as needed to report any concerns.				
Garbage Pickup/Removal	Call for pickups as needed during restoration period.				
Food Operations	Feed employees per schedule for duration of restoration process. Restock supplies as needed.				
Refrigerator	If no electricity, open only when needed. When electricity restored, set controls to normal position				
TV and Cable Systems	Repair any damage				
Fire Alarm System	Repair all damage. Replace locks on shut-off valves. Notify monitoring contractor when repairs are complete.				
Security System	Guards monitor buildings and report any irregularities.				
Storm Supply Storage	Facilitate smooth transition of storm material from staging area to area where needed				
Miscellaneous					
Critical Documents	Bring back once building is dry and secure				
Insurance	Take video of interior and exterior				
Inventory List	Replenish used stock after hurricane.				
Contractor Avaliability Li	Call contractors to assign tasks				

Figure A-4. Post-Storm Activities List

Administrative Issues – Annual Audit

Administrative Issues	YES	NO	N/A	Comments
Injury Recording				
Are we keeping accurate OSHA injury and illness records?				
Are we displaying the OSHA 300A summary from February through April?				
Are we reviewing our injury and illness log for trends on an annual basis?				
Insurance Issues				
Do we have all the policies needed (wind/flood/storm/contents/liability)?				
What are our deductibles and are they at proper levels?				
How will the property be valued (replacement/actual/other)?				
Are we familiar with the riders, addendums, and exclusions?				
Have we performed an annual policy review?				
Am we covered for spoilage, lost income, etc.?				
If we have a lost income policy, what is the duration?				
In what way does the provider require proof of loss?				
What additional loss documentation is needed, and is it stored in a safe place?				
Does it make sense to self-insure?				
Has the insurance coverage been updated to reflect changes in changes in use, building size, or value?				

Administrative Issues – Annual Audit

	YES	NO	NA	Comments
Labor Issues				
Have procedures been developed for tracking emergency-related expenses?				
Do employees know how to contact their families in the event of an emergency?				
Do employees know how to contact their supervisors in the event of an emergency?				
Tenant/Owner-Related Issues				
Who is responsible for staging evacuation drills?				
Is each tenant's business responsible for their own emergency plan?				
Who makes the decision to return to the building after an evacuation?				
What happens to the lease if the property is uninhabitable for an extended period of time?				
What kinds of insurance policies are required and in what amounts?				
Is the landlord, tenant, or both parties required to acquire insurance?				
How much damage must be assessed before the property is considered uninhabitable by the insurance provider?				
How would all occupants be notified?				
Has a chain of command been specified?				
Have all tenants been trained in emergency evacuation procedures?				
What kinds of flammable or toxic chemicals are the tenants storing?				
Do the tenants have all current MSDS documentation?				
Have emergency responsibilities and expectations been added to the lease agreement?				
Who is responsible for alarm testing, fire extinguishers, and other related equipment?				
Does the emergency planner have after-hours phone numbers for the tenants?				
If tenants make changes to the building structure or floor space, have the changes been added to the as-built drawings?				

Emergency Plan Issues – Annual Audit

Emergency Plan Issues	YES	NO	N/A	Comments
Business Continuity				
Have you performed a risk analysis of each of your buildings?				
Have you identified mission critical functions within your organization?				
Have you established a threshold for how many days or hours your business could be down before essential services must be outsourced or moved to a new location?				
Is current data storage process adequate?				
Is redundancy available for electronic data?				
Is off-site storage of vital information being used?				
Are copies of floorplan and emergency action plan available off site?				
Are alternative worksites available?				
Has a process been established for alternate worksite activation?				
Training				
Have members of the emergency action team and supporting employees been adequately trained for their duties?				
Have annual refresher training classes been conducted for emergency response team members?				
Have tabletop exercises been conducted in the last year? When?				
Are certain employees trained in CPR? Who?				
Are certain employees trained in AED usage? Who?				
Are certain employees trained in First aid? Who?				
Are certain employees trained in portable fire extinguisher use? Who?				
Have the employees been trained in adequate emergency evacuation procedures?				

Emergency Plan Issues – Annual Audit

Emergency Plan Questions	YES	NO	N/A	Comments
Does your emergency action plan address the following events:				
Fire				
Drought				
Earthquake				
Hurricane				
Tornado				
Tidal Wave				
Blizzard				
Avalanche				
Tsunami				
Volcano				
Loss of utility				
Loss of communication				
Arson				
Water leak				
Chemical spill				
Acts of war				
Riot or demonstration				
Bomb threat				
Medical emergency				
Workplace violence				
Nuclear accident				
Other				
Does the organization have a formal plan for training its employees in emergency procedures?				
Is training being completed?				
Has your organization met with local community and governmental agencies and briefed them on your plan?				
Do you have updated photos or videotape of the facility and its assets?				
Has the emergency plan been updated to reflect changes in use, building size, or value?				

Emergency Plan Issues – Annual Audit

	YES	NO	N/A	Comments
Does the plan reflect changes in the building layout?				
Does the plan reflect changes in policies and procedures?				
Does this plan identify the person responsible for annual plan update and review?				
Are internal (employees) contact phone numbers accurate?				
Are external (contractors/vendors) contact phone numbers accurate?				
Are suppliers' contact phone numbers accurate?				
Are insurance agents' phone numbers correct?				
Are public safety agencies' phone numbers correct?				
Does the plan reflect lessons learned from the drills and actual emergency events?				
Are problem areas being addressed?				
Are the responsibilities of the following emergency team members assigned:				
Floor captains?				
Fire Brigade?				
Evacuation assistants?				
Other				

Evacuation

	YES	NO	N/A	Comments
Are employees responsible for emergency shut down procedures assigned and capable?				
Who has been assigned to call an actual evacuation?				
What is the preferred means of reporting fires and other emergencies?				
Do you have immediately available emergency scripts that can be read during an emergency event?				

Emergency Plan Issues – Annual Audit

	YES	NO	N/A	Comments
Do you have emergency scripts (or recordings) in languages other than English?				
Can non-English speaking employees understand evacuation orders?				
When was the last three times emergency evacuation drills were conducted?				
Are all shifts covered during emergency evacuation drills?				
If evacuation drills are not practical, do you conduct tabletop exercises?				
Where are the areas of refuge for people with disabilities?				
How will people who remain in the areas of refuge communicate their presence?				
Are evacuation signs currently in place?				
Is the evacuation route clearly marked?				
Are emergency exits clearly marked?				
What are the procedures to account for all employees after an evacuation has been completed?				
Are external staging areas clearly understood?				
What are the rescue and medical duties for those employees assigned to perform them?				
What are the names or job titles of employees who can further explain the emergency action plan?				
Do you have adequate audit forms for post-emergency assessments?				

Emergency Plan Issues – Annual Audit

	YES	NO	N/A	Comments
Evacuation – ADA Issues Have you identified employees with physical or mental limitations that may affect their ability to evacuate the building?				
Have you made accommodations for employees with disabilities during an evacuation?				
Have plans been made to safely evacuate those in wheelchairs or people who have other mobility-related disabilities?				
Have those employees with disabilities been paired with an evacuation assistant?				
Has the assisting partner been trained on how to react and assist people with disabilities?				
Is the assisting partner physically appropriate for the task?				
Have you assigned and trained a backup assisting partner for each disabled partner?				
Is a system established to communicate absences or "not founds" to floor captains?				
Is a system established for disabled partners who do not use their assisting partners (and safely exit the building) to communicate their status to the exterior group leader?				

General Building Issues – Annual Audit

General Building Issues

Alarm

	YES	NO	N/A	Comments
What forms of communication are available during an emergency (i.e. radio, cell phone, public address system, etc.)				
Can audible alarms be heard over machinery?				
Can audible alarms been heard in all locations?				
Is a visual warning system provided for people with hearing disabilities?				
Do you use distinct audible alarms for different emergencies?				
If yes, have employees been trained and do they understand each audible alarm?				
Has the emergency notification system (i.e., alarm and siren) been tested? When?				
Are there an adequate number of alarm pull boxes?				
Are they properly identified and placed in strategic locations?				
Are the pull boxes accessible by all employees, including those with disabilities?				
In locations that feature raised flooring, are water and smoke detectors utilized?				
Is the type of fire extinguisher applicable to the type of fire likely to occur?				
Is the placement of portable fire extinguishers accessible to all employees including those with disabilities?				
Does this building include a fire sprinkler system?				
Has the fire sprinkler system been tested? When? How often?				

Building Issues

	YES	NO	N/A	Comments
Are mechanical rooms kept locked at all times?				
Are electrical rooms locked and accessible by authorized personnel only?				

General Building Issues – Annual Audit

	YES	NO	N/A	Comments
Are tel/com rooms locked and accessible by authorized personnel only?				
Are computer/server rooms locked and accessible by authorized personnel only?				
Is good housekeeping observed in these areas?				
Are computer server rooms segregated from water lines, condensate pans, valves, caps, or plugs?				
Is a moisture detection system used in critical areas?				
Is a swipe card and mag-lock system used to restrict and record entry?				
Do you have a readily available list of the physical addresses of your buildings?				
Do you have the GPS coordinates for your buildings?				
Can emergency response personnel access your building after hours?				
If an external key box exists, has its presence been communicated to emergency response personnel?				
Is a process in place for assigning keys to building occupants?				
Does a key box exist with spare keys?				
Is the key box access limited and safe?				
Are the keys in the key box accurate?				
Is a master key available? Who has it?				
Are shut-off valves properly identified?				
Are shut-off valves labeled for on/off directions?				
Do the shut-off valves require a special key or wrench?				
If yes, where is the key kept?				
Are fire extinguishers inspected and recharged annually or as frequently as required by your local jurisdiction?				

General Building Issues – Annual Audit

	YES	NO	N/A	Comments
If standpipes, fire sprinklers, and hoses exist, are they inspected at least annually?				
Do breaker panels have latching doors?				
Are all breaker panels properly labeled?				
Is adequate space cleared around breaker panel?				
Are you using extension cords rather than permanent wiring?				
Are receptacles and power strips overloaded?				
Are photographs and/or video of the property updated?				
Do the elevators have automatic first floor recall when the fire alarm is tripped?				
Is a process in place to check for and rescue passengers trapped in elevators?				
Is the building address clearly visible from the street?				
Is there sufficient clearance for emergency vehicles beneath overhangs, carports, and parking garages?				
Is the building located near navigable waters that could be used to transport hazardous chemicals?				
Is the building located near railways that could be used to transport hazardous chemicals?				
Is the building located near highways that could be used to transport hazardous chemicals?				
Is the building located near businesses or plants that could experience a release of hazardous chemicals?				
Is the building located within 50 miles of a water dam?				
Are aisles and exit routes of sufficient width for emergency evacuation, including the evacuation of people with disabilities?				

General Building Issues – Annual Audit

	YES	NO	N/A	Comments
Chemical Storage				
Are hazardous chemicals stored properly?				
Are all containers used to store chemicals properly labeled?				
Is MSDS documentation available for all chemicals?				
How often is the MSDS log book updated?				
Are obsolete and superceded MSDS documents archived for 30 years?				
Have employees been trained to properly handle and store chemicals?				
Have employees been trained to recognize and properly handle chemical spills?				
Are chemical spill kits available? If yes, where are they located?				
Has a procedure been established to communicate a spill to employees, management and, if necessary, the proper local agencies?				
Have safe evacuation procedures been established and communicated to employees?				
Emergency Inventory				
Has the emergency food inventory list been reviewed and is it accurate?				
Is the stock of water and food adequate (three days minimum)?				
Are water and food supplies stocked and rotated to ensure freshness?				
Are other emergency supplies such as batteries, first aid, and weather radios stocked?				
Has the emergency equipment inventory list been reviewed and is it accurate?				

General Building Issues – Annual Audit

	YES	NO	N/A	Comments
Equipment Maintenance Is regularly scheduled preventative maintenance being performed on the following equipment:				
Emergency Generator?				
HVAC?				
Cooling Tower?				
Uninterruptible Power Supply?				
Other?				
Equipment Maintenance: Generator Issues If your generator does not include an automatic transfer switch, who will be responsible for switching generator power on?				
Who will confirm the generator power has transferred properly?				
Who will the building occupants call in the event power does not transfer properly or the generator does not start?				
Who will provide regularly scheduled maintenance on the generator?				
Is the generator exhaust near the building's fresh air intake?				
Has the emergency generator been tested under full load? When?				
Has the emergency generator been exercised recently? When? How often?				
Are load shedding priorities established?				
How long can the emergency generator supply power?				
What type of fuel does the emergency generator use?				
Has the generator fuel been treated to prevent spoilage or contamination?				
Who is responsible for refueling the emergency generator?				

General Building Issues – Annual Audit

	YES	NO	N/A	Comments
How will you know if the emergency generator is in need of fuel?				
Do you have a secondary source of fuel for the emergency generator?				
How often is preventative maintenance performed on the emergency generator?				
Are preventative maintenance records for the emergency generator available?				
What equipment is connected to the emergency generator circuit?				
Is the correct (most critical) equipment connected to the generator circuit?				
Is the generator under/over loaded?				
Equipment Maintenance: Other				
How long will the emergency UPS system provide backup power?				
What equipment is connected to the emergency UPS system?				
Can HVAC equipment be shut down from a single location?				
Where are the HVAC fresh air intakes located?				
Is access to HVAC fresh air intakes limited?				
Are they free from possible contamination?				
Are you able to close down the fresh air intakes in an emergency? How?				
Is the mechanical room free of all paper, debris, chemicals, and other storage?				
Are roof drains and gutters kept free of debris?				
Are parking lot drains kept free of debris and operational?				
Are French drains working properly?				
Are sump pumps available and operational?				
Are sand bags and sand available? Can they be quickly shipped?				

General Building Issues – Annual Audit

	YES	NO	N/A	Comments
Floor Plan Are the following items located on the floor plan:				
Windows and Doors				
Designated escape routes				
Emergency exits				
Evacuation staging areas				
Stairways				
Elevators				
Restricted areas				
Mechanical Rooms				
Telecommunications Rooms				
Haz-Mat storage locations				
Loading Dock				
MSDS Station				
Right-to-know station				
Alarm panels				
Fire alarm pull stations				
First aid kits and stations				
Automated External Defibrillators (AED)				
Fire Pumps				
Water sprinkler testing valves				
Fire extinguishers				
Smoke detectors				
Generator location				
Generator powered circuits				
Uninterruptible power system location				
Gas mains and valves				
Water lines and valves				
Utility shutoffs				
Storm drains				
Roof drains				
Sewer lines				

General Building Issues – Annual Audit

	YES	NO	N/A	Comments

General Inspection

Is there any evidence of water leakage or smoke?

Is there any evidence of tampering with access doors, windows, or gate operators?

Is there any evidence of tampering with the security fence?

Are employees at risk of violence or abduction?

Are heavy items placed lower on storage racks, especially in areas susceptible to earthquakes?

Means of Egress

Are all exits doors working properly?

Do all exit doors open outward?

Are exit doors self closing?

Do emergency exit doors have panic hardware installed?

Are all emergency exits doors properly labeled?

Are doors that appear to be exit doors but are not labeled "Not an Exit"?

Are all exit doors kept unlocked and free of obstruction?

If a fire exit door empties into a parking lot area, is the pavement area striped and labeled "No Parking"?

Does the stairwell used as means of egress have working emergency lighting? When was it last checked?

Is the stairwell clean and free of dust and water?

Are fire escape routes clearly identified so that even visitors could recognize and follow them?

General Building Issues – Annual Audit

	YES	NO	N/A	Comments
Are floorplans posted that identify current location and two exit routes?				
At elevators, are signs present that warn of the dangers of use during a fire?				
Are stairwell floors labeled with floor numbers?				
Are floor numbers displayed on the exterior of emergency exit doors in stairwells?				
Are alternate exit routes used during evacuation drills?				

Health Related Issues - Annual Audit

Health-related Issues

	YES	NO	N/A	Comments

AED Issues

How many AEDs do you have installed?

Where are they located?

How will our organization keep informed of changes to regulations involving AED usage?

Where is the prescription from a state-licensed physician for AED installation kept?

Who is trained in proper AED use?

Bloodborne Pathogens

Is the company's "Exposure Control Plan" reviewed and updated annually?

Are employees involved in the selection process of needles and sharps?

Is a "Sharps Injury Log" maintained?

Specific Threat – Annual Audit

Specific Threats	YES	NO	N/A	Comments

Bomb Threat

Are employees provided with a bomb threat checklist?

Have employees been trained in how to evacuate the building in a calm and orderly fashion?

Is a bomb squad available to assist and in what ways will they respond?

Has a bomb search team been established and trained?

Is a set of floor plans available for the search team and authorities?

Has a procedure been established to allow employees back into the building?

Civil Unrest

Do we foresee any events that may cause civil unrest?

Trigger events may include:

- Labor disputes
- Layoffs and downsizing
- Environmentally sensitive meetings or conferences
- Sporting events
- Political rallies
- Economic conferences
- Judicial decisions
- Music concerts
- Religious gatherings
- Biased racial or cultural events

Fire

Have we completed a list of all the major workplace fire hazards and their proper handling and storage requirements?

Have we identified potential ignition sources (i.e., welding, smoking)?

The names of personnel responsible for maintenance of equipment installed to prevent or control fires include:

Specific Threat – Annual Audit

	YES	NO	N/A	Comments
Weather: Heat/Drought				
Have employees been trained in recognizing the dangers of heat and techniques for coping with heat?				
Are heat index warning announced to employees working in hot environments?				
Are mandatory water restrictions in effect (i.e., irrigating landscape)?				
If yes, have we changed automatic timer setting to comply with restrictions?				
In times of drought, have we trimmed back vegetation that could catch fire?				
Have proper precautions been taken when welding or roofing work is performed?				
Weather: Winter Storm Issues				
Are snow removal tools available (shovels, salt, etc.)?				
Have we created a rotating schedule for maintenance employees to be on call after hours in the event of an emergency?				
Do we winterize HVAC, fire sprinkler, and irrigation systems in the fall?				
Have roof drains been checked for blockages that could impede water runoff?				
Have roof gutter systems been checked for blockages that could impede water runoff?				
In the event of a blizzard, do we have a process in place for allowing workers to leave early?				
What will determine if we allow workers to leave early?				
Who will stay behind to monitor the building and perform shutdown if necessary?				

#1 ROOM SEARCH-STOP, LISTEN

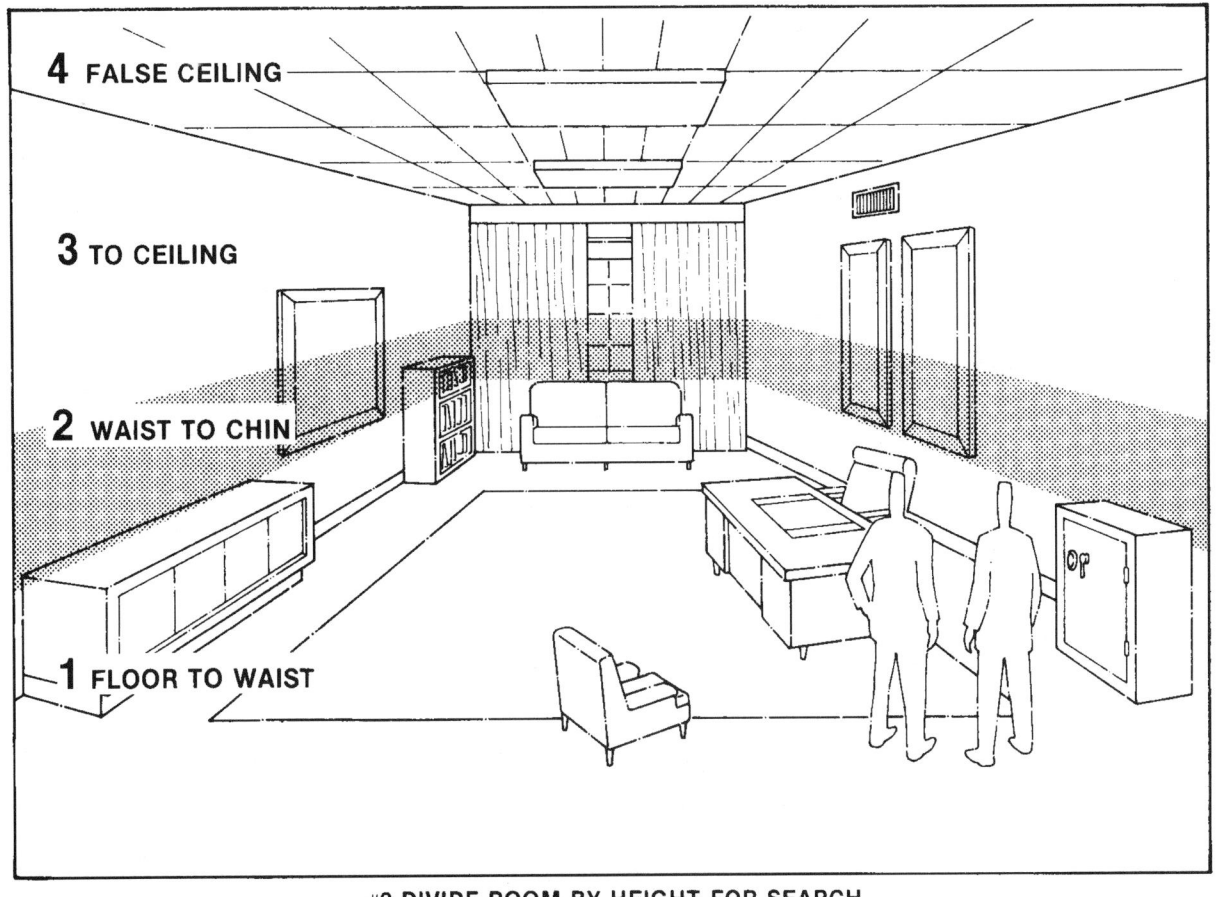

4 FALSE CEILING
3 TO CEILING
2 WAIST TO CHIN
1 FLOOR TO WAIST

#2 DIVIDE ROOM BY HEIGHT FOR SEARCH

Figure A-5. ATF Search and Sweep Procedures

#3 SEARCH ROOM BY HEIGHT & ASSIGNED AREA, OVERLAP FOR BETTER COVERAGE

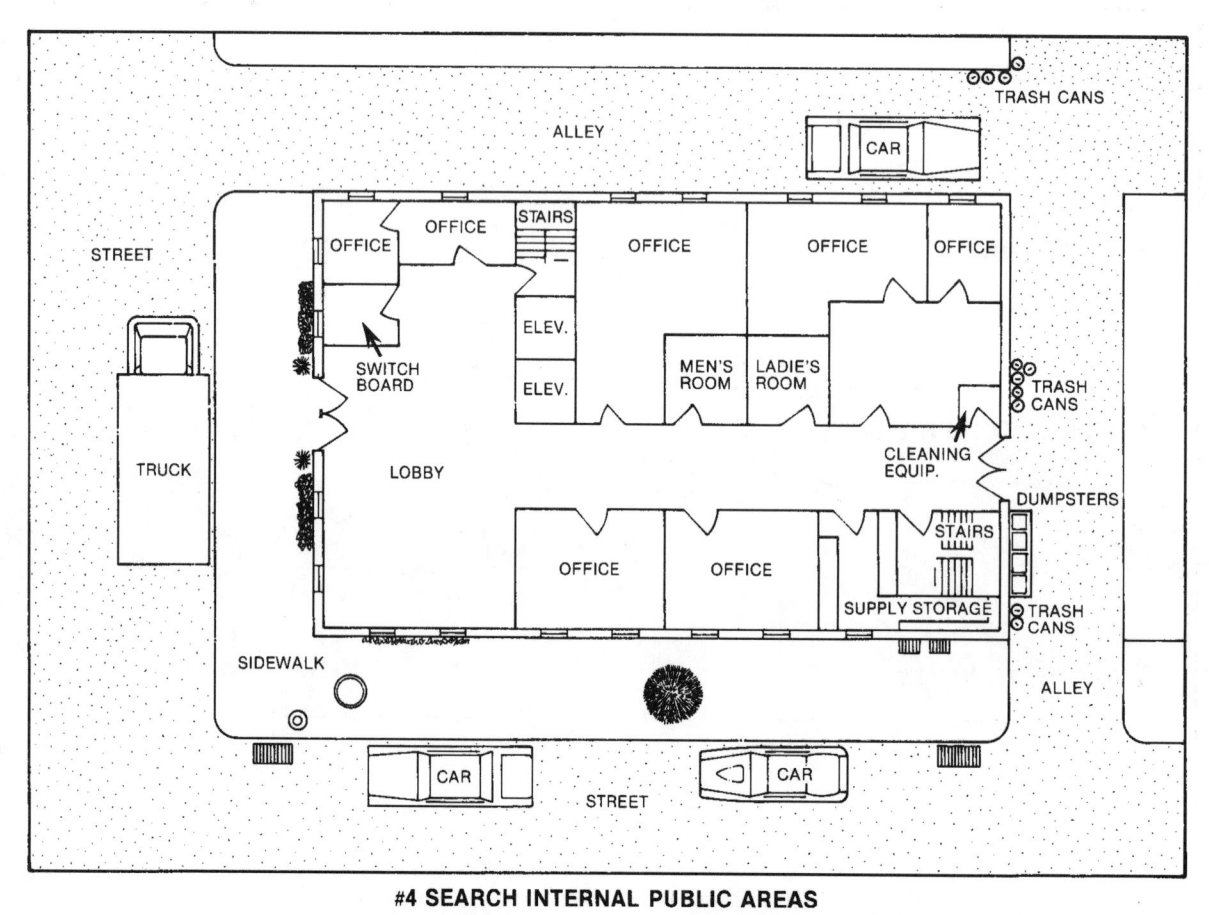

#4 SEARCH INTERNAL PUBLIC AREAS

Figure A-5. ATF Search and Sweep Procedures, continued

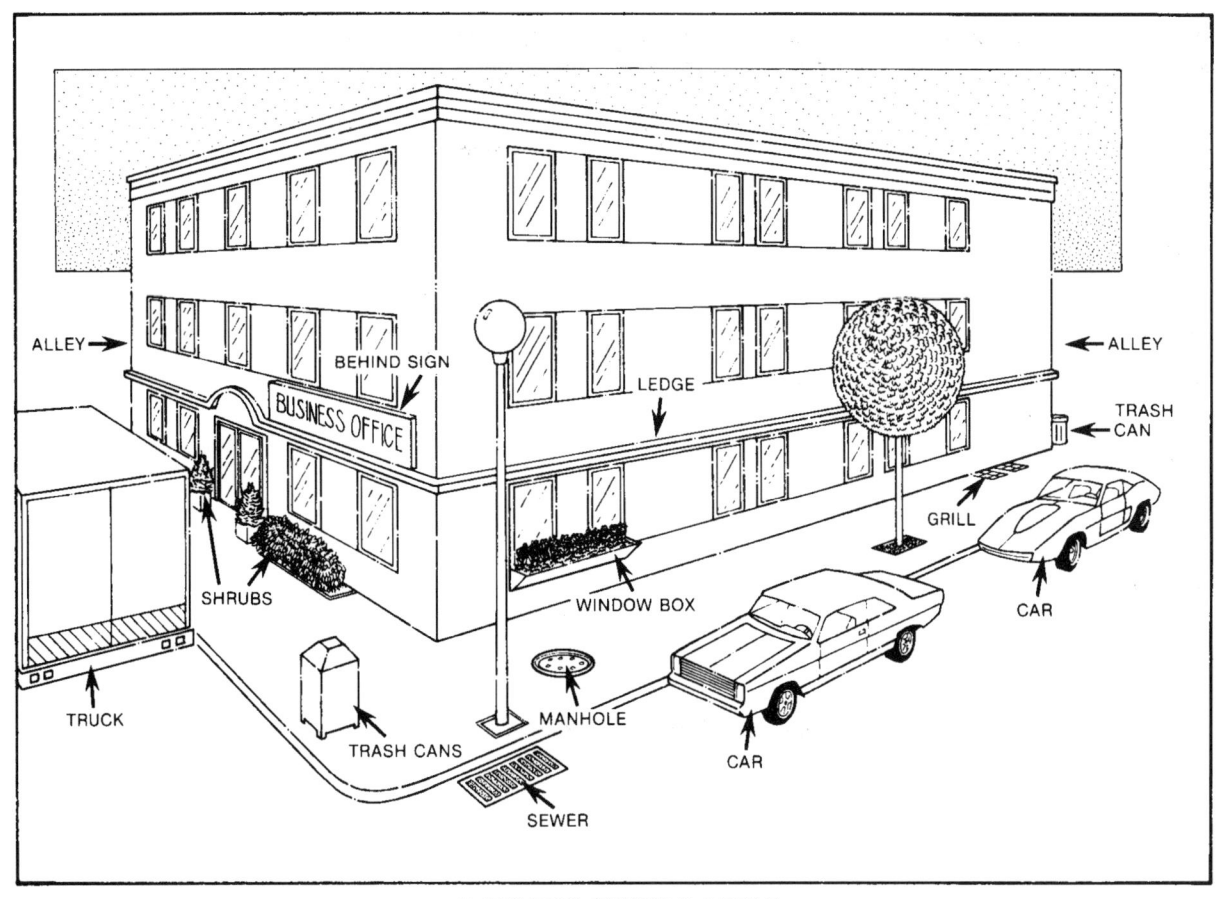

#5 SEARCH OUTSIDE AREAS

Figure A-5. ATF Search and Sweep Procedures, continued

Type of Emergency	Probability	Human Impact	Property Impact	Business Impact	Internal Resource	External Resource	Total
	Hi Lo 5 ←→ 1	Hi Lo 5 ←→ 1	Hi Lo 5 ←→ 1	Hi Lo 5 ←→ 1	Weak 5 ←	Strong → 1	
Fires							
Severe weather							
Hazardous material spills							
Transportation accidents							
Earthquakes							
Terrorism							
Utility outages							
Other							

Figure A-6. Risk Assessment Matrix

Corresponence Log

Date: _____ **Time:** _____

Caller	Issue	Name	Office Phone	Cell Phone	Action Required	Assigned To
☐ Vendor ☐ Contractor ☐ Employee ☐ Police ☐ Fire / Paramedic ☐ Insurance Agent ☐ Other _____					☐ Yes ☐ No	

Date: _____ **Time:** _____

Caller	Issue	Name	Office Phone	Cell Phone	Action Required	Assigned To
☐ Vendor ☐ Contractor ☐ Employee ☐ Police ☐ Fire / Paramedic ☐ Insurance Agent ☐ Other _____					☐ Yes ☐ No	

Date: _____ **Time:** _____

Caller	Issue	Name	Office Phone	Cell Phone	Action Required	Assigned To
☐ Vendor ☐ Contractor ☐ Employee ☐ Police ☐ Fire / Paramedic ☐ Insurance Agent ☐ Other _____					☐ Yes ☐ No	

Figure A-7. Correspndence Log

OSHA
Forms for Recording
Work-Related Injuries and Illnesses

What's Inside...

In this package, you'll find everything you need to complete OSHA's *Log* and the *Summary of Work-Related Injuries and Illnesses* for the next several years. On the following pages, you'll find:

▶ **An Overview: Recording Work-Related Injuries and Illnesses** — General instructions for filling out the forms in this package and definitions of terms you should use when you classify your cases as injuries or illnesses.

▶ **How to Fill Out the Log** — An example to guide you in filling out the *Log* properly.

▶ **Log of Work-Related Injuries and Illnesses** — Several pages of the *Log* (but you may make as many copies of the *Log* as you need.) Notice that the *Log* is separate from the *Summary*.

▶ **Summary of Work-Related Injuries and Illnesses** — Removable *Summary* pages for easy posting at the end of the year. Note that you post the *Summary* only, not the *Log*.

▶ **Worksheet to Help You Fill Out the Summary** — a worksheet for figuring the average number of employees who worked for your establishment and the total number of hours worked.

▶ **OSHA's 301: Injury and Illness Incident Report** — Several copies of the OSHA 301 to provide details about the incident. You may make as many copies as you need or use an equivalent form.

Take a few minutes to review this package. If you have any questions, *visit us online at www.osha.gov* or *call your local OSHA office*. We'll be happy to help you.

U.S. Department of Labor
Occupational Safety and Health Administration

Figure A-8. OSHA 300 Forms and Instructions

An Overview:
Recording Work-Related Injuries and Illnesses

The Occupational Safety and Health (OSH) Act of 1970 requires certain employers to prepare and maintain records of work-related injuries and illnesses. Use these definitions when you classify cases on the Log. OSHA's recordkeeping regulation (see 29 CFR Part 1904) provides more information about the definitions below.

The *Log of Work-Related Injuries and Illnesses* (Form 300) is used to classify work-related injuries and illnesses and to note the extent and severity of each case. When an incident occurs, use the *Log* to record specific details about what happened and how it happened. The *Summary* — a separate form (Form 300A) — shows the totals for the year in each category. At the end of the year, post the *Summary* in a visible location so that your employees are aware of the injuries and illnesses occurring in their workplace.

Employers must keep a *Log* for each establishment or site. If you have more than one establishment, you must keep a separate *Log* and *Summary* for each physical location that is expected to be in operation for one year or longer.

Note that your employees have the right to review your injury and illness records. For more information, see 29 Code of Federal Regulations Part 1904.35, *Employee Involvement*.

Cases listed on the *Log of Work-Related Injuries and Illnesses* are not necessarily eligible for workers' compensation or other insurance benefits. Listing a case on the *Log* does not mean that the employer or worker was at fault or that an OSHA standard was violated.

When is an injury or illness considered work-related?

An injury or illness is considered work-related if an event or exposure in the work environment caused or contributed to the condition or significantly aggravated a preexisting condition. Work-relatedness is presumed for injuries and illnesses resulting from events or exposures occurring in the workplace, unless an exception specifically applies. See 29 CFR Part 1904.5(b)(2) for the exceptions. The work environment includes the establishment and other locations where one or more employees are working or are present as a condition of their employment. See 29 CFR Part 1904.5(b)(1).

Which work-related injuries and illnesses should you record?

Record those work-related injuries and illnesses that result in:
- death,
- loss of consciousness,
- days away from work,
- restricted work activity or job transfer, or
- medical treatment beyond first aid.

You must also record work-related injuries and illnesses that are significant (as defined below) or meet any of the additional criteria listed below.

You must record any significant work-related injury or illness that is diagnosed by a physician or other licensed health care professional. You must record any work-related case involving cancer, chronic irreversible disease, a fractured or cracked bone, or a punctured eardrum. See 29 CFR 1904.7.

What are the additional criteria?

You must record the following conditions when they are work-related:
- any needlestick injury or cut from a sharp object that is contaminated with another person's blood or other potentially infectious material;
- any case requiring an employee to be medically removed under the requirements of an OSHA health standard;
- any Standard Threshold Shift (STS) in hearing (i.e., cases involving an average hearing loss of 10 dB or more in either ear); and
- tuberculosis infection as evidenced by a positive skin test or diagnosis by a physician or other licensed health care professional after exposure to a known case of active tuberculosis.

What is medical treatment?

Medical treatment includes managing and caring for a patient for the purpose of combating disease or disorder. The following are not considered medical treatments and are NOT recordable:
- visits to a doctor or health care professional solely for observation or counseling;
- diagnostic procedures, including administering prescription medications that are used solely for diagnostic purposes; and
- any procedure that can be labeled first aid. (See below for more information about first aid.)

---

1. Within 7 calendar days after you receive information about a case, decide if the case is recordable under the OSHA recordkeeping requirements.

2. Determine whether the incident is a new case or a recurrence of an existing one.

3. Establish whether the case was work-related.

4. If the case is recordable, decide which form you will fill out as the injury and illness incident report.

   You may use *OSHA's 301: Injury and Illness Incident Report* or an equivalent form. Some state workers compensation, insurance, or other reports may be acceptable substitutes, as long as they provide the same information as the OSHA 301.

How to work with the Log

1. Identify the employee involved unless it is a privacy concern case as described below.

2. Identify when and where the case occurred.

3. Describe the case, as specifically as you can.

4. Classify the seriousness of the case by recording the **most serious outcome** associated with the case, with column J (Other recordable cases) being the least serious and column G (Death) being the most serious.

5. Identify whether the case is an injury or illness. If the case is an injury, check the injury category. If the case is an illness, check the appropriate illness category.

U.S. Department of Labor
Occupational Safety and Health Administration

Figure A-8. OSHA 300 Forms and Instructions, continued

What is first aid?

If the incident required only the following types of treatment, consider it first aid. Do NOT record the case if it involves only:

- using non-prescription medications at non-prescription strength;
- administering tetanus immunizations;
- cleaning, flushing, or soaking wounds on the skin surface;
- using wound coverings, such as bandages, BandAids™, gauze pads, etc., or using SteriStrips™ or butterfly bandages.
- using hot or cold therapy;
- using any totally non-rigid means of support, such as elastic bandages, wraps, non-rigid back belts, etc.;
- using temporary immobilization devices while transporting an accident victim (splints, slings, neck collars, or back boards).
- drilling a fingernail or toenail to relieve pressure, or draining fluids from blisters;
- using eye patches;
- using simple irrigation or a cotton swab to remove foreign bodies not embedded in or adhered to the eye;
- using irrigation, tweezers, cotton swab or other simple means to remove splinters or foreign material from areas other than the eye;
- using finger guards;
- using massages;
- drinking fluids to relieve heat stress

How do you decide if the case involved restricted work?

Restricted work activity occurs when, as the result of a work-related injury or illness, an employer or health care professional keeps, or recommends keeping, an employee from doing the routine functions of his or her job or from working the full workday that the employee would have been scheduled to work before the injury or illness occurred.

How do you count the number of days of restricted work activity or the number of days away from work?

Count the number of calendar days the employee was on restricted work activity or was away from work as a result of the recordable injury or illness. Do not count the day on which the injury or illness occurred in this number. Begin counting days from the day after the incident occurs. If a single injury or illness involved both days away from work and days of restricted work activity, enter the total number of days for each. You may stop counting days of restricted work activity or days away from work once the total of either or the combination of both reaches 180 days.

Under what circumstances should you NOT enter the employee's name on the OSHA Form 300?

You must consider the following types of injuries or illnesses to be privacy concern cases:

- an injury or illness to an intimate body part or to the reproductive system,
- an injury or illness resulting from a sexual assault,
- a mental illness,
- a case of HIV infection, hepatitis, or tuberculosis,
- a needlestick injury or cut from a sharp object that is contaminated with blood or other potentially infectious material (see 29 CFR Part 1904.8 for definition), and
- other illnesses, if the employee independently and voluntarily requests that his or her name not be entered on the log. Musculoskeletal disorders (MSDs) are not considered privacy concern cases.

You must not enter the employee's name on the OSHA 300 Log for these cases. Instead, enter "privacy case" in the space normally used for the employee's name. You must keep a separate, confidential list of the case numbers and employee names for the establishment's privacy concern cases so that you can update the cases and provide information to the government if asked to do so.

If you have a reasonable basis to believe that information describing the privacy concern case may be personally identifiable even though the employee's name has been omitted, you may use discretion in describing the injury or illness on both the OSHA 300 and 301 forms. You must enter enough information to identify the cause of the incident and the general severity of the injury or illness, but you do not need to include details of an intimate or private nature.

What if the outcome changes after you record the case?

If the outcome or extent of an injury or illness changes after you have recorded the case, simply draw a line through the original entry or, if you wish, delete or white-out the original entry. Then write the new entry where it belongs. Remember, you need to record the most serious outcome for each case.

Classifying injuries

An injury is any wound or damage to the body resulting from an event in the work environment.

Examples: Cut, puncture, laceration, abrasion, fracture, bruise, contusion, chipped tooth, amputation, insect bite, electrocution, or a thermal, chemical, electrical, or radiation burn. Sprain and strain injuries to muscles, joints, and connective tissues are classified as injuries when they result from a slip, trip, fall or other similar accidents.

Classifying illnesses

Musculoskeletal disorders (MSD illnesses)
A musculoskeletal disorder (MSD) is a disorder of the muscles, nerves, tendons, ligaments, joints, cartilage, or spinal discs. MSDs DO NOT include disorders caused by a slip, trip, motor vehicle accident, fall, or other similar accidents.

Examples of MSD's include:
Carpal tunnel syndrome, epicondylitis, rotator cuff syndrome, tendinitis, De Quervains' disease, Raynaud's phenomenon, trigger finger,

Figure A-8. OSHA 300 Forms and Instructions, continued

U.S. Department of Labor
Occupational Safety and Health Administration

carpet layers knee, tarsal tunnel syndrome, herniated spinal disc, sciatica, and low back pain.

Skin diseases or disorders
Skin diseases or disorders are illnesses involving the worker's skin that are caused by work exposure to chemicals, plants, or other substances.

Examples: Contact dermatitis, eczema, or rash caused by primary irritants and sensitizers or poisonous plants; oil acne; friction blisters, chrome ulcers; inflammation of the skin.

Respiratory conditions
Respiratory conditions are illnesses associated with breathing hazardous biological agents, chemicals, dust, gases, vapors, or fumes at work.

Examples: Silicosis, asbestosis, pneumonitis, pharyngitis, rhinitis or acute congestion; farmer's lung, beryllium disease, tuberculosis, occupational asthma, reactive airways dysfunction syndrome (RADS), chronic obstructive pulmonary disease (COPD), hypersensitivity pneumonitis, toxic inhalation injury, such as metal fume fever, chronic obstructive bronchitis, and other pneumoconioses.

Poisoning
Poisoning includes disorders evidenced by abnormal concentrations of toxic substances in blood, other tissues, other bodily fluids, or the breath that are caused by the ingestion or absorption of toxic substances into the body.

Examples: Poisoning by lead, mercury, cadmium, arsenic, or other metals; poisoning by carbon monoxide, hydrogen sulfide, or other gases; poisoning by benzene, benzol, carbon tetrachloride, or other organic solvents; poisoning by insecticide sprays, such as parathion or lead arsenate; poisoning by other chemicals, such as formaldehyde.

Noise-Induced Hearing loss
Noise-induced hearing loss is defined for recordkeeping purposes as a change in hearing threshold relative to the baseline audiogram of an average of 10 dB or more in either ear at 2000, 3000 and 4000 hertz.

All other Illnesses
All other occupational illnesses.

Examples: Heatstroke, sunstroke, heat exhaustion, heat stress and other effects of environmental heat; freezing, frostbite, and other effects of exposure to low temperatures; decompression sickness; effects of ionizing radiation (isotopes, x-rays, radium); effects of nonionizing radiation (welding flash, ultra-violet rays, lasers); anthrax; bloodborne pathogenic diseases, such as AIDS, HIV, hepatitis B or hepatitis C; brucellosis; malignant or benign tumors; histoplasmosis; coccidioidomycosis.

When must you post the Summary?
You must post the *Summary* only — not the *Log* — by February 1 of the year following the year covered by the form and keep it posted until April 30 of that year.

How long must you keep the Log and Summary on file?
You must keep the *Log* and *Summary* for 5 years following the year to which they pertain.

Do you have to send these forms to OSHA at the end of the year?
No. You do not have to send the completed forms to OSHA unless specifically asked to do so.

How can we help you?
If you have a question about how to fill out the *Log*,

- ☐ **visit us online at www.osha.gov** or
- ☐ **call your local OSHA office.**

U.S. Department of Labor
Occupational Safety and Health Administration

Figure A-8. OSHA 300 Forms and Instructions, continued

Optional

Calculating Injury and Illness Incidence Rates

What is an incidence rate?

An incidence rate is the number of recordable injuries and illnesses occurring among a given number of full-time workers (usually 100 full-time workers) over a given period of time (usually one year). To evaluate your firm's injury and illness experience over time or to compare your firm's experience with that of your industry as a whole, you need to compute your incidence rate. Because a specific number of workers and a specific period of time are involved, these rates can help you identify problems in your workplace and/or progress you may have made in preventing work-related injuries and illnesses.

How do you calculate an incidence rate?

You can compute an occupational injury and illness incidence rate for all recordable cases or for cases that involved days away from work for your firm quickly and easily. The formula requires that you follow instructions in paragraph (a) below for the total recordable cases or those in paragraph (b) for cases that involved days away from work, and for both rates the instructions in paragraph (c).

(a) *To find out the total number of recordable injuries and illnesses that occurred during the year*, count the number of line entries on your OSHA Form 300, or refer to the OSHA Form 300A and sum the entries for columns (G), (H), (I), and (J).

(b) *To find out the number of injuries and illnesses that involved days away from work*, count the number of line entries on your OSHA Form 300 that received a check mark in column (H), or refer to the entry for column (H) on the OSHA Form 300A.

(c) *The number of hours all employees actually worked during the year*. Refer to OSHA Form 300A and optional worksheet to calculate this number.

You can compute the incidence rate for all recordable cases of injuries and illnesses using the following formula:

Total number of injuries and illnesses ÷ Number of hours worked by all employees x 200,000 hours = Total recordable case rate

(The 200,000 figure in the formula represents the number of hours 100 employees working 40 hours per week, 50 weeks per year would work, and provides the standard base for calculating incidence rates.)

You can compute the incidence rate for recordable cases involving days away from work using the following formula:

The number of injuries and illnesses that involved days away from work ÷ Number of hours worked by all employees x 200,000 hours = Incidence rate for cases involving days away from work

You can use the same formula to calculate incidence rates for other variables such as cases involving restricted work activity (column (I) on Form 300A), cases involving musculoskeletal disorders (column (M-2) on Form 300A), etc. Just substitute the appropriate total for these cases, from Form 300A, into the formula in place of the total number of injuries and illnesses.

What can I compare my incidence rate to?

The Bureau of Labor Statistics (BLS) conducts a survey of occupational injuries and illnesses each year and publishes incidence rate data by various classifications (e.g., by industry, by employer size, etc.). You can obtain these published data at www.bls.gov or by calling a BLS Regional Office.

Worksheet

Total number of recordable injuries and illnesses in your establishment

[ ] ÷ [ ] X 200,000 = [ ] Total recordable cases incidence rate

Hours worked by all your employees

Total number of recordable injuries and illnesses with days away from work

[ ] ÷ [ ] X 200,000 = [ ] Cases involving days away from work incidence rate

Hours worked by all your employees

Figure A-8. OSHA 300 Forms and Instructions, continued

U.S. Department of Labor
Occupational Safety and Health Administration

How to Fill Out the Log

The *Log of Work-Related Injuries and Illnesses* is used to classify work-related injuries and illnesses and to note the extent and severity of each case. When an incident occurs, use the *Log* to record specific details about what happened and how it happened.

If your company has more than one establishment or site, you must keep separate records for each physical location that is expected to remain in operation for one year or longer.

We have given you several copies of the *Log* in this package. If you need more than we provided, you may photocopy and use as many as you need.

The *Summary* — a separate form — shows the work-related injury and illness totals for the year in each category. At the end of the year, count the number of incidents in each category and transfer the totals from the *Log* to the *Summary*. Then post the *Summary* in a visible location so that your employees are aware of injuries and illnesses occurring in their workplace.

You don't post the Log. You post only the Summary at the end of the year.

Identify the person

Describe the case

Be as specific as possible. You can use two lines if you need more room.

Revise the log if the injury or illness progresses and the outcome is more serious than you originally recorded for the case. Cross out, erase, or white-out the original entry.

Classify the case

Choose ONE of these categories. Classify the case by recording the most serious outcome of the case, with column J (Other recordable cases) being the least serious and column G (Death) being the most serious.

Note whether the case involves an injury or an illness.

U.S. Department of Labor
Occupational Safety and Health Administration

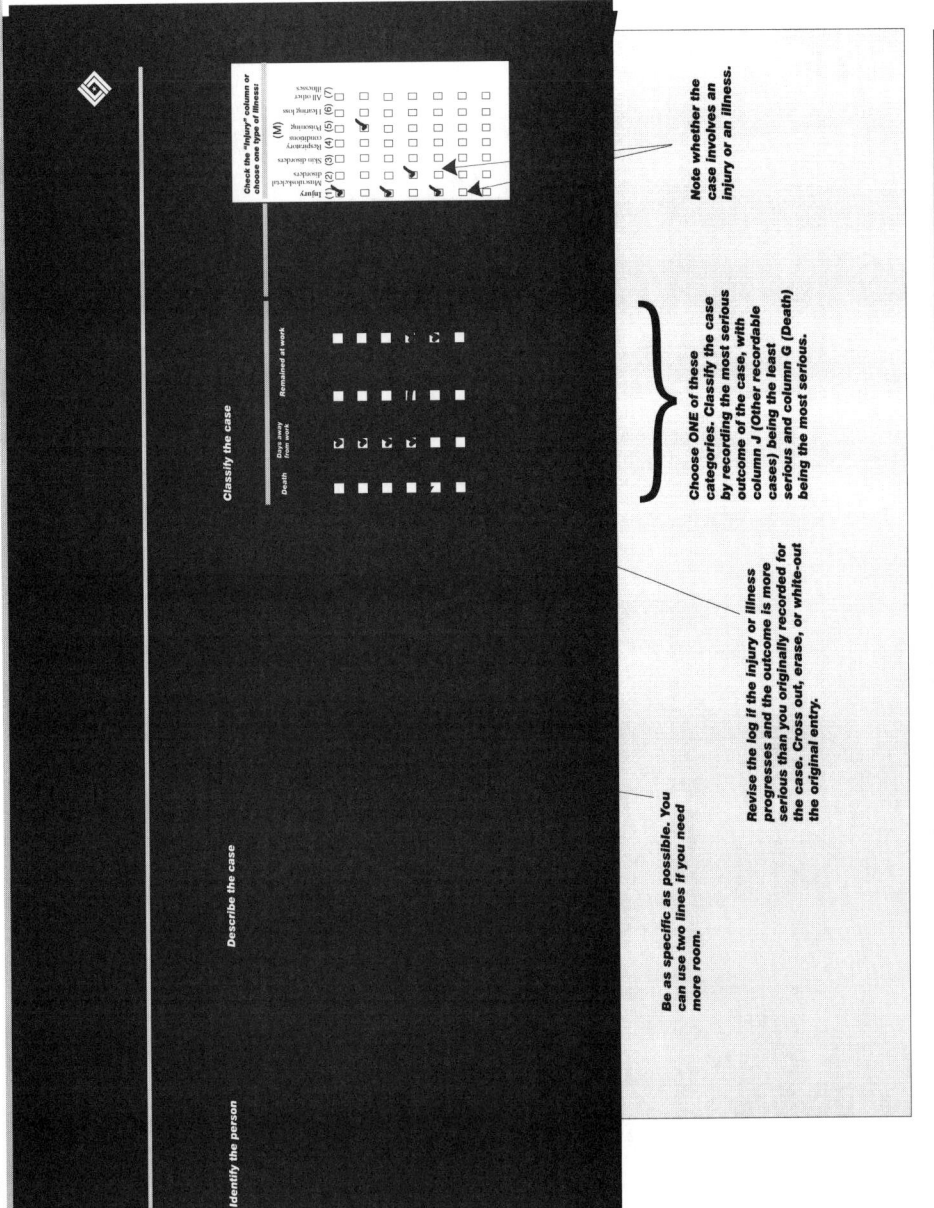

Figure A-8. OSHA 300 Forms and Instructions, continued

OSHA's Form 300

Log of Work-Related Injuries and Illnesses

Attention: This form contains information relating to employee health and must be used in a manner that protects the confidentiality of employees to the extent possible while the information is being used for occupational safety and health purposes.

U.S. Department of Labor
Occupational Safety and Health Administration

Year 20____

Form approved OMB no. 1218-0176

You must record information about every work-related death and about every work-related injury or illness that involves loss of consciousness, restricted work activity or job transfer, days away from work, or medical treatment beyond first aid. You must also record significant work-related injuries and illnesses that are diagnosed by a physician or licensed health care professional. You must also record work-related injuries and illnesses that meet any of the specific recording criteria listed in 29 CFR Part 1904.8 through 1904.12. Feel free to use two lines for a single case if you need to. You must complete an Injury and Illness Incident Report (OSHA Form 301) or equivalent form for each injury or illness recorded on this form. If you're not sure whether a case is recordable, call your local OSHA office for help.

Establishment name _____

City _____ State _____

| (A) Case no. | (B) Employee's name | (C) Job title (e.g., Welder) | (D) Date of injury or onset of illness | (E) Where the event occurred (e.g., Loading dock north end) | (F) Describe injury or illness, parts of body affected, and object/substance that directly injured or made person ill (e.g., Second degree burns on right forearm from acetylene torch) | Using these four categories, check ONLY the most serious result for each case: Death (G) / Days away from work (H) / Job transfer or restriction (I) / Other recordable cases (J) | Enter the number of days the injured or ill worker was: On job transfer or restriction (K) / Away from work (L) | Check the "Injury" column or choose one type of illness: (M) Injury (1) / Skin disorder (2) / Respiratory condition (3) / Poisoning (4) / Hearing loss (5) / All other illnesses (6) |

Page totals ▶

Be sure to transfer these totals to the Summary page (Form 300A) before you post it.

Public reporting burden for this collection of information is estimated to average 14 minutes per response, including time to review the instructions, search and gather the data needed, and complete and review the collection of information. Persons are not required to respond to the collection of information unless it displays a currently valid OMB control number. If you have any comments about these estimates or any other aspects of this data collection, contact: US Department of Labor, OSHA Office of Statistics, Room N-3644, 200 Constitution Avenue, NW, Washington, DC 20210. Do not send the completed forms to this office.

Page ____ of ____

Figure A-8. OSHA 300 Forms and Instructions, continued

Using the Log, count the individual entries you made for each category. Then write the totals below, making sure you've added the entries from every page of the Log. If you had no cases, write "0."

Employees, former employees, and their representatives have the right to review the OSHA Form 300 in its entirety. They also have limited access to the OSHA Form 301 or its equivalent. See 29 CFR Part 1904.35, in OSHA's recordkeeping rule, for further details on the access provisions for these forms.

Number of Cases

Total number of deaths

_____
(G)

Total number of cases with days away from work

_____
(H)

Total number of cases with job transfer or restriction

_____
(I)

Total number of other recordable cases

_____
(J)

Number of Days

Total number of days of job transfer or restriction

_____
(K)

Total number of days away from work

_____
(L)

Injury and Illness Types

Total number of . . .
(M)
(1) Injuries _____
(2) Musculoskeletal disorders _____
(3) Skin disorders _____
(4) Respiratory conditions _____
(5) Poisonings _____
(6) Hearing loss cases _____
(7) All other illnesses _____

Establishment Information

Your establishment name _____
Street _____
City _____ State _____ ZIP _____

Industry description (e.g., Manufacture of motor truck trailers)
_____

Standard Industrial Classification (SIC), if known (e.g., SIC 3715)
_____

Employment Information (If you don't have these figures, see the Worksheet on the back of this page to estimate.)

Annual average number of employees _____
Total hours worked by all employees last year _____

Sign here

Knowingly falsifying this document may result in a fine.

I certify that I have examined this document and that to the best of my knowledge the entries are true, accurate, and complete.

_____ _____
Company executive   Title

( ) ___ - ___   __/__/__
Phone   Date

Post this Summary page from February 1 to April 30 of the year following the year covered by the form.

Public reporting burden for this collection of information is estimated to average 50 minutes per response, including time to review the instructions, search and gather the data needed, and complete and review the collection of information. Persons are not required to respond to the collection of information unless it displays a currently valid OMB control number. If you have any comments about these estimates or any other aspects of this data collection, contact: US Department of Labor, OSHA Office of Statistics, Room N-3644, 200 Constitution Avenue, NW, Washington, DC 20210. Do not send the completed forms to this office.

Figure A-8. OSHA 300 Forms and Instructions, continued

Optional

Worksheet to Help You Fill Out the Summary

At the end of the year, OSHA requires you to enter the average number of employees and the total hours worked by your employees on the summary. If you don't have these figures, you can use the information on this page to estimate the numbers you will need to enter on the Summary page at the end of the year.

How to figure the average number of employees who worked for your establishment during the year:

① **Add** the total number of employees your establishment paid in all pay periods during the year. Include all employees: full-time, part-time, temporary, seasonal, salaried, and hourly.

The number of employees paid in all pay periods = _____ ①

② **Count** the number of pay periods your establishment had during the year. Be sure to include any pay periods when you had no employees.

The number of pay periods during the year = _____ ②

③ **Divide** the number of employees by the number of pay periods.

$\dfrac{①}{②}$ = _____ ③

④ **Round the answer** to the next highest whole number. Write the rounded number in the blank marked *Annual average number of employees.*

The number rounded = _____ ④

For example, Acme Construction figured its average employment this way:

For pay period...	Acme paid this number of employees...
1	10
2	0
3	15
4	30
5 ▶	40
...	20
24	20
25	15
26	+10
	830

Number of employees paid = 830
Number of pay periods = 26

$\dfrac{830}{26}$ = 31.92

31.92 rounds to 32

32 is the annual average number of employees

How to figure the total hours worked by all employees:

Include hours worked by salaried, hourly, part-time and seasonal workers, as well as hours worked by other workers subject to day to day supervision by your establishment (e.g., temporary help services workers).

Do not include vacation, sick leave, holidays, or any other non-work time, even if employees were paid for it. If your establishment keeps records of only the hours paid or if you have employees who are not paid by the hour, please estimate the hours that the employees actually worked.

If this number isn't available, you can use this optional worksheet to estimate it.

Optional Worksheet

Find the number of full-time employees in your establishment for the year.

_____

Multiply by the number of work hours for a full-time employee in a year.

X _____

This is the number of full-time hours worked.

_____

Add the number of any overtime hours as well as the hours worked by other employees (part-time, temporary, seasonal)

+ _____

Round the answer to the next highest whole number. Write the rounded number in the blank marked *Total hours worked by all employees last year.*

_____

U.S. Department of Labor
Occupational Safety and Health Administration

Figure A-8. OSHA 300 Forms and Instructions, continued

OSHA's Form 301
Injury and Illness Incident Report

Attention: This form contains information relating to employee health and must be used in a manner that protects the confidentiality of employees to the extent possible while the information is being used for occupational safety and health purposes.

U.S. Department of Labor
Occupational Safety and Health Administration

Form approved OMB no. 1218-0176

This *Injury and Illness Incident Report* is one of the first forms you must fill out when a recordable work-related injury or illness has occurred. Together with the *Log of Work-Related Injuries and Illnesses* and the accompanying *Summary*, these forms help the employer and OSHA develop a picture of the extent and severity of work-related incidents.

Within 7 calendar days after you receive information that a recordable work-related injury or illness has occurred, you must fill out this form or an equivalent. Some state workers' compensation, insurance, or other reports may be acceptable substitutes. To be considered an equivalent form, any substitute must contain all the information asked for on this form.

According to Public Law 91-596 and 29 CFR 1904, OSHA's recordkeeping rule, you must keep this form on file for 5 years following the year to which it pertains.

If you need additional copies of this form, you may photocopy and use as many as you need.

Completed by _____
Title _____
Phone (___) ___ - ____   Date ___/___/___

Information about the employee

1) Full name _____
2) Street _____
   City _____ State ____ ZIP _____
3) Date of birth ___/___/___
4) Date hired ___/___/___
5) ☐ Male ☐ Female

Information about the physician or other health care professional

6) Name of physician or other health care professional _____
7) If treatment was given away from the worksite, where was it given?
   Facility _____
   Street _____
   City _____ State ____ ZIP _____
8) Was employee treated in an emergency room?
   ☐ Yes ☐ No
9) Was employee hospitalized overnight as an in-patient?
   ☐ Yes ☐ No

Information about the case

10) Case number from the Log _____ *(Transfer the case number from the Log after you record the case.)*
11) Date of injury or illness ___/___/___
12) Time employee began work _____ AM / PM
13) Time of event _____ AM / PM ☐ Check if time cannot be determined
14) **What was the employee doing just before the incident occurred?** Describe the activity, as well as the tools, equipment, or material the employee was using. Be specific. *Examples:* "climbing a ladder while carrying roofing materials"; "spraying chlorine from hand sprayer"; "daily computer key-entry."
15) **What happened?** Tell us how the injury occurred. *Examples:* "When ladder slipped on wet floor, worker fell 20 feet"; "Worker was sprayed with chlorine when gasket broke during replacement"; "Worker developed soreness in wrist over time."
16) **What was the injury or illness?** Tell us the part of the body that was affected and how it was affected; be more specific than "hurt," "pain," or sore." *Examples:* "strained back"; "chemical burn, hand"; "carpal tunnel syndrome."
17) **What object or substance directly harmed the employee?** *Examples:* "concrete floor"; "chlorine"; "radial arm saw." *If this question does not apply to the incident, leave it blank.*
18) **If the employee died, when did death occur?** Date of death ___/___/___

Public reporting burden for this collection of information is estimated to average 22 minutes per response, including time for reviewing instructions, searching existing data sources, gathering and maintaining the data needed, and completing and reviewing the collection of information. Persons are not required to respond to the collection of information unless it displays a current valid OMB control number. If you have any comments about this estimate or any other aspects of this data collection, including suggestions for reducing this burden, contact: US Department of Labor, OSHA Office of Statistics, Room N-3644, 200 Constitution Avenue, NW, Washington, DC 20210. Do not send the completed forms to this office.

Figure A-8. OSHA 300 Forms and Instructions, continued

If You Need Help...

If you need help deciding whether a case is recordable, or if you have questions about the information in this package, feel free to contact us. We'll gladly answer any questions you have.

▼ **Visit us online at www.osha.gov**

▼ **Call your OSHA Regional office and ask for the recordkeeping coordinator**

or

▼ **Call your State Plan office**

Federal Jurisdiction

Region 1 - 617 / 565-9860
Connecticut; Massachusetts; Maine; New Hampshire; Rhode Island

Region 2 - 212 / 337-2378
New York; New Jersey

Region 3 - 215 / 596-1201
DC; Delaware; Pennsylvania; West Virginia

Region 4 - 404 / 562-2300
Alabama; Florida; Georgia; Mississippi

Region 5 - 312 / 353-2220
Illinois; Ohio; Wisconsin

Region 6 - 214 / 767-4731
Arkansas; Louisiana; Oklahoma; Texas

Region 7 - 816 / 426-5861
Kansas; Missouri; Nebraska

Region 8 - 303 / 844-1600
Colorado; Montana; North Dakota; South Dakota

Region 9 - 415 / 975-4310

Region 10 - 206 / 533-5930
Idaho

State Plan States

Alaska - 907 / 269-4957

Arizona - 602 / 542-5795

California - 415 / 703-5100

*Connecticut - 860 / 566-4380

Hawaii - 808 / 586-9100

Indiana - 317 / 232-3325

Iowa - 515 / 281-3606

Kentucky - 502 / 564-3070

Maryland - 410 / 767-2371

Michigan - 517 / 322-1851

Minnesota - 651 / 296-2116

Nevada - 702 / 687-3250

*New Jersey - 609 / 292-2313

New Mexico - 505 / 827-4230

*New York - 518 / 457-2574

North Carolina - 919 / 807-2875

Oregon - 503 / 378-3272

Puerto Rico - 787 / 754-2171

South Carolina - 803 / 734-9632

Tennessee - 615 / 741-2793

Utah - 801 / 530-6901

Vermont - 802 / 828-2765

Virginia - 804 / 786-6613

Virgin Islands - 340 / 772-1315

Washington - 360 / 902-5554

Wyoming - 307 / 777-7786

*Public Sector only

U.S. Department of Labor
Occupational Safety and Health Administration

Figure A-8. *OSHA 300 Forms and Instructions, continued*

Have questions?

If you need help in filling out the *Log* or *Summary*, or if you have questions about whether a case is recordable, contact us. We'll be happy to help you. You can:

▶ Visit us online at: **www.osha.gov**

▶ Call your regional or state plan office. You'll find the phone number listed inside this cover.

U.S. Department of Labor
Occupational Safety and Health Administration

Figure A-8. OSHA 300 Forms and Instructions, continued

Helpful Websites

American Red Cross	<www.redcross.org>
Avalanches	<www.avalanche.org>
Disaster Center	<www.disastercenter.com>
Drought	<www.usgs.gov>
Earthquake history	<earthquake.usgs.gov/>
Emergency Management Training	<www.carlyleconsultants.com>
Federal Emergency Management Agency (FEMA)	<www.fema.gov>
Historical Hurricane Tracks	<www.nhc.noaa.gov/pastall.html>
Landslide Risk Areas	<landslides.usgs.gov>
National Hurricane Center	<www.nhc.noaa.gov>
National Oceanographic & Atmospheric Administration (NOAA)	<www.noaa.gov>
National Weather Service	<www.nws.noaa.gov>
Natural Hazards Statistics	<www.nws.noaa.gov/om/hazstats.shtm>
State Dam Safety Officials	<www.damsafety.org>
Tornado Project Online	<www.tornadoproject.com>
Tsunami Warning Center	<www.tsunami.gov>
U.S. Army Corps of Engineers	<corpsgeo1.usace.army.mil>
Volcano hazards	<volcanoes.usgs.gov>
Wildfire Dangers	<www.fs.fed.us/land/wfas/fd_class.gif>
Wildland Fire Updates	<www.nifc.gov/fireinfo/nfn.html>
Wind Zones	<www.esri.com/hazards/>

Where Else Can I Find Assistance?

Creating an emergency action plan is quite an involved exercise. If you have never been through this exercise before, you may easily become confused or frustrated. Of course this book will help your planning efforts tremendously. Additionally, a number of other resources exist and you should take advantage of their expertise. Mid- to large-sized businesses should meet with local governmental agencies and community organizations. By simply informing them of your efforts to create an emergency plan and asking them for their assistance, you will tap into an incredible resource. Often, they will be willing to assist you by coordinating planning, training or mitigation efforts. They may be willing to share their plan. Since they've been through the planning experience, they will be able to add valuable insight and information. Don't hesitate to contact these outside organizations.

Local Fire Department

Your local fire department is an excellent resource for assistance and training. Typically, your local fire department will work with you in planning and executing your emergency evacuation drills. They may be willing to review your emergency action plan and offer suggestions. Quite often they will provide specific safety-related training such as proper fire extinguisher use. Well versed in all aspects of emergency planning and recovery, fire departments are an excellent resource and should not be overlooked.

Utility Companies

Utility companies play a critical role during times of emergency. In scenarios in which electrical power is lost for an extended period of time, a utility company can play a role in the speed of your organization's recovery. Certain businesses such as hospitals, fire rescue, and telecommunications facilities are critical to the community and must be made operational as quickly as possible. If your business provides a critical service, it would be wise to contact the local utility company and identify your critical needs. They may have some flexibility with when they will restore electrical power to your business. Remember, however, restoration of service will not be accelerated based on your perceived need. Instead, any early restoration will be based on your organization's impact on the community's ability to quickly recover from the emergency event.

American Red Cross

The American Red Cross is a humanitarian organization led by volunteers. The Red Cross has been instrumental in providing relief to victims of all types of disasters. The Red Cross also assists people in preparing for, responding to, and preventing emergencies. The Red Cross has provided health- and safety-related training for over a century. These education programs include such diverse topics such as first aid, AED, CPR, HIV/AIDS, swimming, and life guarding programs. In the average year, nearly 12 million people take Red Cross heath and safety courses.

Although not a government agency, the Red Cross works closely with various government agencies including the Federal Emergency Management Agency (FEMA) during times of major crises. To learn more about the Red Cross, including the training and education programs available in your area, visit their web site at <www.redcross.org>

Safety Consultants

Many organizations do not have in-house expertise when it comes to emergency planning, training, and recovery. For those organizations, consultants specializing in emergency planning and recovery can be invaluable.

The Carlyle Consulting Group is a consulting firm specializing in the corporate and commercial real estate arena. The consulting service and training workshops focus on easy-to-implement and common sense solutions to today's complex business problems including OSHA compliance and emergency action planning, training, and development. For a complete listing of services provides, visit the Carlyle Consulting Group's website.

Carlyle Consulting Group
1225 S. Lakeside Drive
Lake Worth, Florida 33463
Office: (561) 533-0153
Toll Free: (866) 227-5953
Fax: (561) 547-0232
<www.carlyleconsultants.com>

In 1862, American Bureau of Shipping (ABS) began its legacy of safety in the maritime industry. Today ABS Consulting, Inc., has broadened its focus far beyond the marine industry with services relevant to virtually every market segment today. ABS Consulting identifies hazards, analyzes risk, and provides mitigation services to leading organizations worldwide. From structural engineering to risk consulting to catastrophe management, ABS Consulting offers solutions tailored to meet the specific needs of its clients.

The Government Institute Division of ABS Consulting specializes in public and private training and publishing for professionals in environmental management, occupational safety and health, shipping, food and drug, energy, risk management, and quality. Government Institutes' authors and course presenters are recognized authorities in their disciplines.

ABS Consulting, Government Institutes Division
4 Research Place
Suite 200
Rockville, MD 20850 U.S.A.
Tel (301) 921-2300
Fax (301) 921-0373
<www.absconsulting.com>

Federal Emergency Management Agency (FEMA)

The Federal Emergency Management Agency (FEMA) is an independent government agency reporting to the President. FEMA's mission is to plan for, respond to, and prevent disasters across the country. FEMA employs over 2,500 emergency response personnel, supplemented by over 5,000 standby-by reservists.

The services of FEMA are very diverse and include emergency planning, training, and coordination of response and the administration of assistance to states, communities, businesses, and individuals. Furthermore, FEMA is instrumental in flood plain management and administering the national flood insurance programs.

FEMA
500 C Street
SW Washington, D.C. 20472
Phone: (202) 566-1600
<www.fema.gov>

United States Fire Administration

The United States Fire Administration (USFA) is an entity of the Federal Emergency Management Agency (FEMA). Its mission is to save lives and reduce economic losses due to fire and fire-related emergencies. The United States Fire Administration provides training and education services to the emergency services community. They have been very influential in developing fire safety–related codes for states and municipalities.

USFA
16825 South Seton Avenue
Emmitsburg, MD 21727
Phone (301) 447-1000
Fax   (301) 447-1052
<www.usfa.fema.gov>

National Emergency Management Association (NEMA)

The National Emergency Management Association (NEMA) is a professional association of emergency management directors committed to providing expertise and leadership in the field of emergency planning and management. NEMA also takes a leadership role in developing strategic partnerships that provide continuous improvements in emergency management functions. The NEMA website is an excellent resource for up-to-date information on topics such emergency preparedness, including homeland security and bioterrorism.

National Emergency Management Association
c/o Council of State Governments
P.O. Box 11910

Lexington, KY 40578
(859) 244-8000
(859) 244-8239
<www.nemaweb.org>

National Fire Protection Association (NFPA)

The National Fire Protection Association (NFPA) is recognized as the leading authority where fire safety–related education, training, and code development is concerned. Municipalities across the nation have adopted NFPA standards as a basis for their building codes. The NFPA is an international nonprofit organization whose mission is to reduce the burden of fire and other related hazards. Its goal is to increase quality of life for all by advocating scientifically sound codes and standards, research, training, and education.

The National Fire Protection Association is comprised of more than 80 national trade and professional organizations with a membership exceeding 75,000 professional worldwide.

NFPA
1 Batterymarch Park
P.O. Box 9101
Quincy, Massachusetts
USA 02269-9101
(617) 770-3000
<www.nfpa.org>

Building Owners and Managers Association (BOMA)

The Building Owners and Managers Association (BOMA) was founded in 1907 under the National Association of Building Owners and Managers name. Today BOMA International offers education, benchmarking studies, building performance data, and related services to real estate professionals who manage over 8.5 billion square feet of office space in North America. BOMA International represents over 100 North American associations and nine international associations.

BOMA International works to increase professionalism through designation curriculum and continuing education programs including those issues dealing with OSHA compliance, emergency planning, and disaster recovery in a commercial real estate environment. To realize that mission, in 1970 the Building Owners and Managers Institute (BOMI) was established by BOMA to provide industry-related curriculum. BOMI has developed four distinct professional designations—the Real Property Administrator (RPA®), the Facilities Management Administrator (FMA®), the Systems Maintenance Administrator (SMA®), and the Systems Maintenance Technician (SMT®) designations. In addition to subjects such as building maintenance and financial management knowledge, the BOMI curriculum teaches real estate professionals about topics such as OSHA compliance and emergency planning and management.

BOMA
1201 New York Avenue, NW
Suite 300
Washington, DC 20005
Phone: 202-408-2662
Fax: 202-371-0181
Fax on Demand: 732-544-2515
<www.boma.org>

International Facility Managers Association (IFMA)

The International Facility Management Association (IFMA) was established in 1980 as a non-profit, worldwide professional association of facility managers. IFMA membership exceeds 18,000 facility professionals in 50 countries. IFMA members make up over 129 chapters worldwide.

IFMA is a leader in providing direct and timely real estate training, including topics on emergency planning in a built environment. In addition to IFMA's regional training programs, World Workplace is an annual professional conference featuring educational programs, extensive networking opportunities, and the largest exposition of products and services relevant to the built environment.

IFMA
1 E. Greenway Plaza, Suite 1100
Houston, TX 77046-0194
Phone: (713) 623-4362
Fax: (713) 623-6124
<www.ifma.org>

Institute of Real Estate Management (IREM)

The Institute of Real Estate Management (IREM) was founded in 1933 and today remains a leading organization of professional real managers and owners. IREM is an affiliate of the National Association of Realtors (NAR), and its mission is to educate real estate managers, certify their competence, and serve as an advocate on issues affecting the real estate management industry. These issues include emergency planning and management in the real estate sector.

Institute of Real Estate Management
430 N. Michigan Ave.
Chicago, IL 60611-4090
Phone: (800) 837-0706

Fax: (800) 338-4736
<www.irem.org>

American Institute of Architects (AIA)

The American Institute of Architects (AIA) serves the needs of registered architects across the nation. The AIA has been instrumental in educating architects in areas of as diverse as building design and building security. As the industry deals with new issues such as terrorism, security, and disabled employees, the AIA provides direction and influences design and code development across the nation.

AIA
1735 New York Ave.
NW Washington, DC 20006
Phone: (800) AIA-3837
Fax: (202) 626-7547
<http://www.aia.org/>

Occupational Safety and Health Administration (OSHA)

In 1970, the Occupational Safety and Health Act was created. The Act gave birth to the Occupational Safety and Health Administration (OSHA), a federal agency whose goal is to save lives, prevent injuries, and protect the health of over 100 million workers in the United States. OSHA employs approximately 2,100 inspectors in addition to engineers, physicians, educators, standards writers, and other technical and support personnel spread over more than 200 offices throughout the country. These OSHA representatives develop safety standards, enforces those standards, and proactively educate both employers and employees about safe work. In the three decades since the inception of OSHA, injuries, illnesses, and deaths related to the work environment have steadily declined.

OSHA is a federal agency that has jurisdiction over half of the states in the union. The remaining states have created their own state OSHA plans, which supercede Federal OSHA. The state plans must be as strict and all encompassing as Federal OSHA, and in some cases the state plans are even more demanding. If you are unsure of your responsibilities, visit the OSHA web site and search for "State Plans."

U.S. Department of Labor
Occupational Safety and Health Administration
200 Constitution Avenue
Washington, D.C. 20210
Phone: (800) 321-OSHA (6742)

Americans with Disabilities Act (ADA)

The Americans with Disabilities Act was created to provide protection from bias and ensure fair and equal treatment to persons with disabilities. This includes protection during times of emergency and building evacuation. The ADA is a federal act and covers all commercial buildings; however, local municipalities may enact codes that are more strict than the federal regulations. The ADA has been instrumental in developing standards for building ingress and egress, resulting in numerous life-saving and life-enhancing laws for the benefit of the disabled.

U.S. Department of Justice
950 Pennsylvania Avenue, NW
Civil Rights Division
Disability Rights Section—NYAVE
Washington, D.C. 20530
Phone: (800) 514-0301
<www.usdoj.gov/crt/ada/adahom1.htm>

Corporate Real Estate Network (CoreNet)

On May 1, 2002, International Development Research Council (IDRC) and The International Association of Corporate Real Estate Executives (NACORE International) joined forces. This merger of two leading and respected associations led to the creation of CoreNet Global.

CoreNet Global consists of 58 chapters in five global regions including Asia, Australia, Europe, Latin America, and North America including Canada. The nearly 7,000 members of corporate real estate executive, service providers, and developers make CoreNet Global the voice of the industry. CoreNet Global has been instrumental in developing best practices for security and emergency planning in the corporate real estate industry.

CoreNet Global
440 Columbia Drive, Suite 100
West Palm Beach, FL 33409
Phone: (800) 726-8111
Fax: (561) 697-4853

The Weather Channel

In 1982 the Weather Channel was born. Although many people found it difficult to believe a 24-hour weather channel could succeed, today, the Weather Channel is the leading weather provider on the Internet and on television. The Weather Channel's website averages more than 130 million page views each month. The Weather Channel provides viewers with up-to-date weather conditions, five-day forecasts, and severe weather alerts. Because weather plays such an important role in emergency events, the

Weather Channel seeks to keep the public informed about developing weather related problems.

<www.weather.com>

National Oceanic and Atmospheric Administration (NOAA)

The National Oceanic and Atmospheric Administration (NOAA) was created in 1970 to serve a national need for better protection of life and property from natural hazards. NOAA conducts research and evaluates changes in the oceans, atmosphere, space, and sun. Based on the data collected, NOAA works to predict and mitigate the effects of these environmental changes.

The National Weather Service (NWS) is a member of the NOAA organization and provides weather-related forecasts for the United States. In addition to weather forecasts, the National Weather Service provides severe weather warnings. The National Weather Service the official voice for issuing warnings during life-threatening weather situations in the United States. Because the nation is affected by weather-related events outside of the United States, the National Weather Services tracks these events as well. Additionally, the NWS website is a great source for Doppler radar, satellite observations, and a detailed glossary of weather-related terms.

National Oceanic and Atmospheric Administration
<www.noaa.gov>

National Weather Service
<www.nws.noaa.gov>

Here are just a few of the companies that manufacture the NOAA weather radio:

First Alert Weather Radios
1212 Remington Road
Schaumburg, IL 60173
Toll Free (800) 259-0959
E-Mail: info@wirelessmarketing.com
<www.cherokeeelectronics.com/weather.html>

Homesafe Inc.
1501 N. Raleigh Street
Suite 5
Highway 55
Angier, NC 27501
Toll Free (800) 607-6737
E-Mail: HSIsupport@aol.com
<www.homesafeinc.com>

Midland Electronics
1690 N. Topping
Kansas City, MO 64120
Phone (816) 241-8500
E-Mail: Midlndcb@midlandradio.com
<www.midlandradio.com>

Radio Shack
300W 3rd. Street
Suite 1400
Ft. Worth, TX 76102
Phone (817) 415-3200
<www.radioshack.com>

WeatherRadios.com
1251 N. Sherwood
Palatine, IL 60067
Toll Free (800) 818-6505
E-Mail: info@weatherradios.com
<www.weatherradios.com>

Index

A

ABS Consulting, 292
Adult learners, 141
AIDS, 167
Atlantic Ocean, 42
Alarms, 126, 230, 261
    Audible, 109, 147
    Strobe, 109
Alaskan tsunami, 42
Alcohol, Tobacco, and Firearms (ATF), 57, 273–275
Aleutian Islands, 43
Alfred P. Murrah Federal Building, 52, 208
Allergies, 106
American Airlines, 53
American Express, 128
American Heart Association, 167, 169
American Institute of Architects (AIA), 296
Americans with Disabilities Act (ADA), 103, 105–106, 152–156, 260, 297
    Areas of refuge, 155
    Partner system, 106
Anthrax, 53, 61–64
Areas of refuge, 155
Asbestos, 184–185
Asbestos containing material, 184
As-builts, 121, 222
Assessments, 173 (see also audits)
Association of State Dam Safety Officials, (ASDSO) 29, 99
Asthma, 106

Atlas, Randall, 106
Atlas Safety & Security Design, 203–204
Atofina Chemicals, 66
Atropine, 65
Audits, 173
    Administrative, 254–255
    Emergency plan, 256–260
    General building, 261–269
    Health related, 270
    Specific threats, 271–272
Aum Shinrikyo cult, 64
Automated External Defibrillator (AED), 122, 162, 166–168, 270
Avalanches, 9–14, 233, 290

B

Base Flood Elevation (BFE), 28
Beverly Hills Supper Club, 23
Bhopal, India, 66
Billion Dollar Weather Disasters, 3
Biological issues, 61–66, 152, 243
Bioterrorism, 61, 65
Blizzards, 9–13, 233, 272
    Blizzard warning, 13
    Great Appalachian Blizzard, 10
    Snow removal, 11
Bloodborne Pathogen Standard, 77, 167, 270
Blueprints, 121, 222
Boilers, 11
Bomb search, 56, 273–275

Bomb threat, 51–61, 241, 271
Bomb threat checklist, 54
Botulism, 63
Broken bones, 166
Brucellosis, 63
Buddy system (see partner system)
Building codes, 207
Building Owners and Management Association, (BOMA) xvii, 294–295
Building Performance Assessment Team (BPAT), 207
Bureau of Alcohol, Tobacco and Firearms, 54, 57
    Bomb threat checklist, 54
Burns
    Chemical, 164
    Electrical, 164
    First degree, 164
    Laser, 164
    Second degree, 164
    Thermal, 164
    Third degree, 164
    Victims, 164
Business continuity, 256
Business interruption, 93

C

California brownouts, 73
California's seven-year drought, 13
Cardiopulmonary Resuscitation (CPR), 162, 167–169
Carlyle Consulting Group, xvii, 290, 292
Casavant, David, xvii
Catering, 182
Center for Disease Control (CDC), 63–65, 243
Chain of command, 92, 97, 157, 217
Chain of craters, 47
Chain of survival, 168
Charleston, South Carolina, 32
Charlotte, North Carolina, 32

Chemical
    Agents, 61–66, 152, 243
    Exposure, 165
    Inventory list, 104
    Spills, 66–70, 97, 243
    Storage, 264
Chicago code, 144
Civil disturbance, 70–72, 245, 271
Cold sites, 128
Colonial Industrial Products, 164
Columbine High School, 83, 85
Command center, 60, 129, 178, 231
Command post leader, 158
Communication, 193–199
Press release, samples, 197–198, 250
Consolidated Edison Electric Company, 73
Consultants, safety, 125, 178, 290, 292
Contingency centers, 128
    Hot sites, 128
    Cold sites, 128
Contractor availability form, 36, 116, 218
Contractors, 115–116
Cooling tower collapse, 80
Core business, 92
Corporate Real Estate Network (CoreNet), 297
Correspondence log, 180, 277
Crime Prevention Through Environmental Design (CPTED) 203–204

D

Dade County, Florida, 31, 33
Data security, 130, 232
Defibrillator (see Automated External Defibrillator)
Denny, Reginald, 70
Disabilities, 105–106, 152–156 (see also Americans with Disabilities Act)
Disaster, definition of, 1
Disaster aid, 6
Disaster center, 290

Disaster Mitigation Act, 116
Doors (see exit doors)
Doppler radar, 5
Drought, 13–23, 233–234, 272, 290
    California's seven year drought, 13

E

Earthquakes, 18, 234–235, 290
    Northridge earthquake, 6, 19
    Loma Prieta, 18
Ebola, 63
Egress, 33, 143, 268 (see also evacuation)
Electric Power Research Institute (EPRI), 74
Elevators, 153
Emergency account number, 179
Emergency action plan, 111–113, 138, 213, 257
Emergency Alert System (EAS), 5
Emergency Broadcast System (EBS), 5
Emergency, definition of, 1
Emergency drills, 37
    Evaluating performance of, 161
    Full scale, 39
    Orientation, 139, 141
    Pre-evacuation, 139
    Tabletop, 139, 141, 224
    Walk-through, 139
Emergency Management Institute (EMI), 209
Emergency management team, 114, 216
Emergency supplies, 118, 219, 264
Emergency survival kit, 21
Emergency training, 137, 256, 290
Enron, 74
Environmental issues, 183
Environmental Protection Agency (EPA), 66, 69
Environmental Systems Research Institute (ESRI), 28, 101
Escape route, 102
Evacuation, 54, 102, 154, 158–160, 225, 258–260
    Announced, 141
    Full, 224
    Partial, 224
    Pre–evacuation, 139
    Script, 225
    Unannounced, 141
Evacuation routes, 33, 143, 268
Exit doors, 156, 204
Exposure Control Plan, 77
External support, 115

F

Facilities, 97–99
    Assessments, 173
    Audits, 173
    Commercial, 97
    Educational, 98, 160
    Healthcare, 98, 147
    Industrial, 97
    Limited access, 201
    Point of entry, 201
    Public access, 202
Federal Emergency Management Agency (FEMA), 6–7, 28, 33, 77, 80–81, 100–101, 207–210, 246, 290
Fire, 23–25, 235–236, 271, 290
    Beverly Hills Supper Club, 23
Fire brigade, 24, 97, 150, 157
Fire extinguisher, 123
Fire prevention code, 207
Fire safety director, 145
Fire suppression systems, 123
Fire sprinkler systems, 12
    Deluge system, 25, 124
    Dry pipe system, 25, 124
    Wet pipe system, 25, 124
Fire wardens, 145
First aid, 162
First aid kit, 162, 227
Flash flood warning, 27, 33, 237
Flash flood watch, 27, 33, 237

Flood, 26–29, 48, 236
    Base Flood Elevation (BFE) 28
Flood warning, 27, 32, 237
Flood watch, 27, 32, 237
Flood zone, 29
Flood Insurance Rate Map (FIRM), 28
Flood Insurance Study (FIA), 28
Floor captains, 50, 157
Florida Keys Hurricane, 31
Florida State University, 72
Fujita categories, 40

G

General Chemical, 68
General Duty Clause, 83–84
General Services Administration (GSA), 205
Generators, 122, 132–136, 229, 265–266
    Diesel, 133
    Gasoline, 134
    Load shedding, 134
    Natural gas, 134
    Transfer switch, 135
Geographic Information System (GIS), 28, 101
Geographic location, 95
Good Samaritan law, 162, 168
Government Institute, 292
Governor Guy Hunt, 14
Great Appalachian Blizzard, 10
Great Chicago Flood, 26
Group leader, 158
Gulf of Mexico, 41

H

Hackers, 130
Harris, Eric, 83
Harrisburg, Pennsylvania, 78
Hazardous chemicals, 95
Hearing impairments, 108
Heat index, 14
Heating, ventilating, and air conditioning (HVAC), 12, 152
Hepatitis B virus, 77, 167
Hepatitis C virus, 77, 167
High-rise building, 144
Hilo, Hawaii, 43
Homestead, Florida, 31
Homicide, 82
Hooker Chemicals, 67
Hot sites, 128
Houston, Texas, 31
Human Immunodeficiency Virus (HIV), 77 (see also AIDS)
Hurricane, 29–38, 237, 290
    Andrew, 31, 33, 174, 178, 191
    Camille, 31
    Georges, 209
    warning, 32, 237
    watch, 32, 237
Hyatt Regency walkway collapse, 81

I

Indemnity agreement, 118
Injury recording, 187–190, 251, 254
Institute of Real Estate Management (IREM), 295–296
Insurance, 173, 196, 223, 254
Internal support, 115
International Facility Managers Association (IFMA), xvii, 295
International Tsunami Information Center, 43–44
Inventory list, 119
Iran–Iraq war, 64

J

Job Accommodation Network, 155
Joint Commission on Accreditation of Healthcare Organizations (JCAHO), 93, 98

K

Kaczynski, Ted, 62
Kilauea volcano, 47
Klebold, Dylan, 83
Knox-Box, 123

L

Lake Pontchartrain, 27
Landslides, 42, 290
Lawn irrigation, 11
Leased property, 94
Lee, John, 178
Liability, 100
Lifting techniques, 155
Load shedding, 134 (see also utility)
Location key, 151 (see also blueprints)
Loma Prieta, 18
Los Angeles Police Department, 70
Los Angeles riots, 70
Love Canal, 67

M

Massey, Curtis, 125
Massey Enterprises, 125
Material Safety Data Sheets (MSDS), 68, 98, 103–105, 165
Mauna Loa, Hawaii, 47
McIlvane, Thomas, 83
Meals Ready to Eat (MRE), 118–120, 182
Media, 194 (see also communication, press releases)
Medical Emergencies, 74–75, 242
    American Heart Association, 167, 169
    Automated External Defibrillator (AED), 122, 162, 166–168, 270
    Cardiopulmonary Resuscitation (CPR), 162, 167–169
Melioidosis, 63
Merrill Lynch, 128
Millennium, 85
Mills, Harry, 90
Mission statement, 90–91, 112, 215
Mount St. Helens, 46 (see also volcanoes)
Mustard gas, 65

N

National Association for Search & Rescue, 81
National Climatic Data Center, 3
National Drought Mitigation Center (NDMC), 17
National Electric Code (NEC), 102
National Emergency Management Association (NEMA), 293
National Fire Alarm Code, 128
National Fire Protection Association (NFPA), 103, 124, 127–128, 294
National Guard, 38, 70
National Hurricane Center, 30, 290
National Institute of Standards and technology (NIST), 153
National Oceanic and Atmospheric Administration (NOAA), 4, 43, 290, 298
National Safety Council, 76
National Weather Service (NWS), 4, 38, 43, 290, 298
Natural events, 1, 9–49, 233
Navigable waterway, 95
Needlestick Safety Prevention Act, 76
Nerve agents, 64
New Orleans, 1, 27
New York Blackout of 1977, 73 (see also utility)
New York Stock Exchange, 131
Niagara Falls, New York, 67
Nipah virus, 64
NOAA weather radio, 4, 37–38, 298–299
Non-natural events, 1, 51–84
Northeast ice storm, 9
Northridge earthquake, 6, 19
Nuclear exposure, 244

Nuclear Regulatory Commission, 77, 244

O

Occupational Safety and Health Agency (OSHA), 24, 67, 76–77, 80, 92, 101–105, 112, 136, 150, 156, 167, 185–190, 296
    Bloodborne Pathogen Standard, 77, 167, 270
  OSHA Forms (300, 300A & 301), 187–190, 251, 278–289
    Sharps Injury Log, 77
Off-site storage, 130, 232
Oklahoma City, 41, 52, 208
Oleum, 68
Online hazard maps, 101
OSHA (see Occupational Safety and Health Agency)

P

Pacific Ocean, 42
Pacific Tsunami Warning Center, 19, 44
Palmer Z index, 14
Partner system, 106
Pearl Harbor, 53
Pentagon, 53, 124, 205
Personal Protective Equipment (PPE), 238, 246
Philadelphia, Pennsylvania, 66
Pittsburg, Pennsylvania, 53
Plague, 63
Planning, 87
  Cost of, 88–89
Point Pleasant, West Virginia, 80
Post–failure analysis, 206
Post–storm activity list, 253
Power plants, nuclear, 78
Pregnancy, 106
Pre-storm activity list, 252
Press release, samples, 197–198, 250
Presumed asbestos containing material, 184
Prince William Sound, 42 (see also tsunami)

Project impact, 28
PS Plus, 164
Public address system, 148

Q

Q fever, 63

R

Red Cross, 33, 76, 169, 291, 290
Resources, 88
Restoration, 171, 175, 248
Richmond, California, 68
Right-to-Know Standard, 104
Ring of fire, 46 (see also volcanoes)
Riot, 71
Risk assessment matrix, 99, 276
Risk Management Plan (RMP), 67
Royal Oak Postal Facility, 83
Runner, 158

S

Saffir-Simpson rating, 30, 238
Salmonella, 63
Salvage, 173
Sarin, 64
Schrader, Gerhard, 64
Self Contained Breathing Apparatus (SCBA), 153
Seattle, Washington, 71
Sharps Injury Log, 77 (see also OSHA)
Signage, 103
Silverstein Properties, 181
Sleeping accommodations, 182
Slidell, Louisiana, 27
Small pox, 63, 65
Snow removal, 11
Soman, 64
South Florida Building Code, 33
Southern heat wave, 13
Special Flood Hazard Areas (SFHA), 28

Staging area, 98, 156–157, 226
Stand-off distance, 205
Storm activity list, 34–35
Storm preparation list, 220–221
Storm shutters, 205
Strobe alarm, 109
Structural collapse, 80–81, 246
Sunscreen, 38
Super Outbreak, 39 (see also tornadoes)
Superfund Amendments and Reauthorization Act (SARA), 67
Sylmar, California, 70

T

Tabun, 64
Tallahassee, Florida, 72
Terrorists, 51–52
Three Mile Island, 78
Transfer switch, 135 (see also generators)
Travel advisory, 13
Tornadoes, 39–42, 238–239, 290
    Fujita categories, 40
    Super Outbreak, 39
    Tornado alley, 41
    Tornado warning, 33, 39, 239
    Tornado watch, 33, 39, 239
Tropical Storm Allison, 31
Tropical storm warning, 33, 237
Tropical storm watch, 33, 237
Tsunamis, 42–45, 239–240, 290
    Alaskan tsunami, 42
    Pacific Tsunami Warning Center, 19, 44
    Prince William Sound, 42
    West Coast Tsunami Warning Center, 19
Tsunami laboratory, 43
Tsunami museum, 43
Two-way radio, 61

U

Unabomber, 62
Uninterruptible Power Supply (UPS), 132
Union Carbide, 66
United Airlines, 53
United States Fire Administration (USFA), 293
Urban Search and Rescue Grid, 125
U.S. Army Corps of Engineers, 290
U.S. Department of Agriculture (USDA), 15
U.S. Department of Interior, 47
U.S. Drought monitor, 16
U.S. Geological Survey, 47
U.S. Postal Service, 62
Utility
    California brownouts, 73
    Loss of, 73–74, 241
    New York Blackout of 1977, 73
    Shut–offs, 122, 151

V

Vendors, 116
Virtual reality software, 206
Visitors, to buildings, 156
Visual impairments, 107
Volcanoes, 42–49, 46, 240–241, 290
    Chain of craters, 47
    Kilauea volcano, 47
    Ring of fire, 46
    Mount St. Helens, 46

W

Weather Channel, 297–298
West Coast Tsunami Warning Center, 19, 44
Western United States Wild Fires, 23
Wheelchair, 154 (see also ADA)
Wind zones, 290
Windows, 204
Winter storm warning, 13
Winter storm watch, 13
Workplace violence, 82–84, 247
    Columbine High School, 83, 85
    Royal Oak Postal Facility, 83

World Trade Center, xv, 51–53, 93, 124, 181, 193, 206
World Trade Organization, 71, 143
World War I, 65
World War II, 65

X

Xenia, Ohio, 39

Y

Y2K, 85
Yellowstone National Park, 13

Z

Zeriscape, 13 (see also drought)
Zero tolerance, 84

Government Institutes Mini-Catalog

PC #	ENVIRONMENTAL TITLES	Pub Date	Price*
629	ABCs of Environmental Regulation	1998	$65
672	Book of Lists for Regulated Hazardous Substances, 9th Edition	1999	$95
4100	CFR Chemical Lists on CD ROM, 1999-2000 Edition	1999	$125
512	Clean Water Handbook, Second Edition	1996	$115
581	EH&S Auditing Made Easy	1997	$95
673	E H & S CFR Training Requirements, Fourth Edition	2000	$99
825	Environmental, Health and Safety Audits, 8th Edition	2001	$115
548	Environmental Engineering and Science	1997	$95
643	Environmental Guide to the Internet, Fourth Edition	1998	$75
820	Environmental Law Handbook, Sixteenth Edition	2001	$99
688	EH&S Dictionary: Official Regulatory Terms, Seventh Edition	2000	$95
821	Environmental Statutes, 2001 Edition	2001	$115
4099	Environmental Statutes on CD ROM for Windows-Single User, 1999 Ed.	1999	$169
707	Federal Facility Environmental Compliance and Enforcement Guide	2000	$115
708	Federal Facility Environmental Management Systems	2000	$99
689	Fundamentals of Site Remediation	2000	$85
515	Industrial Environmental Management: A Practical Approach	1996	$95
510	ISO 14000: Understanding Environmental Standards	1996	$85
551	ISO 14001: An Executive Report	1996	$75
588	International Environmental Auditing	1998	$179
518	Lead Regulation Handbook	1996	$95
608	NEPA Effectiveness: Mastering the Process	1998	$95
582	Recycling & Waste Mgmt Guide to the Internet	1997	$65
615	Risk Management Planning Handbook	1998	$105
603	Superfund Manual, 6th Edition	1997	$129
685	State Environmental Agencies on the Internet	1999	$75
566	TSCA Handbook, Third Edition	1997	$115
534	Wetland Mitigation: Mitigation Banking and Other Strategies	1997	$95

PC #	SAFETY and HEALTH TITLES	Pub Date	Price*
697	Applied Statistics in Occupational Safety and Health	2000	$105
547	Construction Safety Handbook	1996	$95
553	Cumulative Trauma Disorders	1997	$75
663	Forklift Safety, Second Edition	1999	$85
709	Fundamentals of Occupational Safety & Health, Second Edition	2001	$69
612	HAZWOPER Incident Command	1998	$75
662	Machine Guarding Handbook	1999	$75
535	Making Sense of OSHA Compliance	1997	$75
718	OSHA's New Ergonomic Standard	2001	$95
558	PPE Made Easy	1998	$95
683	Product Safety Handbook	2001	$95
598	Project Mgmt for E H & S Professionals	1997	$85
658	Root Cause Analysis	1999	$105
552	Safety & Health in Agriculture, Forestry and Fisheries	1997	$155
669	Safety & Health on the Internet, Third Edition	1999	$75
668	Safety Made Easy, Second Edition	1999	$75
590	Your Company Safety and Health Manual	1997	$95

Government Institutes

4 Research Place, Suite 200 • Rockville, MD 20850-3226
Tel. (301) 921-2323 • FAX (301) 921-0264
Email: giinfo@govinst.com • Internet: http://www.govinst.com

Please call our customer service department at (301) 921-2323 for a free publications catalog.

CFRs now available online. Call (301) 921-2355 for info.

*All prices are subject to change. Please call for current prices and availablity.

Government Institutes Order Form

4 Research Place, Suite 200 • Rockville, MD 20850-3226
Tel (301) 921-2323 • Fax (301) 921-0264
Internet: http://www.govinst.com • E-mail: giinfo@govinst.com

4 EASY WAYS TO ORDER

1. **Tel:** **(301) 921-2323**
   Have your credit card ready when you call.

2. **Fax:** **(301) 921-0264**
   Fax this completed order form with your company purchase order or credit card information.

3. **Mail:** **Government Institutes Division**
   ABS Group Inc.
   P.O. Box 846304
   Dallas, TX 75284-6304   USA
   Mail this completed order form with a check, company purchase order, or credit card information.

4. **Online:** Visit http://www.govinst.com

PAYMENT OPTIONS

❑ **Check** *(payable in US dollars to **ABS Group Inc. Government Institutes Division**)*

❑ **Purchase Order** *(This order form must be attached to your company P.O. Note: All International orders must be prepaid.)*

❑ **Credit Card**   ❑ VISA   ❑ MasterCard   ❑ American Express

Exp. ___ /____

Credit Card No. _____

Signature _____

(Government Institutes' Federal I.D.# is 13-2695912)

CUSTOMER INFORMATION

Ship To: (Please attach your purchase order)

Name _____
GI Account # *(7 digits on mailing label)* _____
Company/Institution _____
Address _____
(Please supply street address for UPS shipping)
_____
City _____ State/Province _____
Zip/Postal Code _____ Country _____
Tel ( ) _____
Fax ( ) _____
E-mail Address _____

Bill To: (if different from ship-to address)

Name _____
Title/Position _____
Company/Institution _____
Address _____
(Please supply street address for UPS shipping)
_____
City _____ State/Province _____
Zip/Postal Code _____ Country _____
Tel ( ) _____
Fax ( ) _____
E-mail Address _____

Qty.	Product Code	Title	Price

Subtotal _____
MD Residents add 5% Sales Tax _____
Shipping and Handling (see box below) _____
Total Payment Enclosed _____

30 DAY MONEY-BACK GUARANTEE
If you're not completely satisfied with any product, return it undamaged within 30 days for a full and immediate refund on the price of the product.

SOURCE CODE: BP03

Shipping and Handling
Within U.S:
1-4 products: $6/product
5 or more: $4/product
Outside U.S:
Add $15 for each item (Global)

Sales Tax
Maryland 5%
Texas 8.25%
Virginia 4.5%